Learn

Eureka Math™

Grade 1
Module 1

Published by Great Minds®.

Copyright © 2018 Great Minds®.

Printed in the U.S.A.
This book may be purchased from the publisher at eureka-math.org.
10 9 8 7 6 5 4 3 2 1
v1.0 PAH
ISBN 978-1-64054-050-7

G1-M1-L-05.2018

Learn ◆ Practice ◆ Succeed

Eureka Math™ student materials for *A Story of Units*® (K–5) are available in the *Learn, Practice, Succeed* trio. This series supports differentiation and remediation while keeping student materials organized and accessible. Educators will find that the *Learn, Practice,* and *Succeed* series also offers coherent—and therefore, more effective—resources for Response to Intervention (RTI), extra practice, and summer learning.

Learn

Eureka Math Learn serves as a student's in-class companion where they show their thinking, share what they know, and watch their knowledge build every day. *Learn* assembles the daily classwork—Application Problems, Exit Tickets, Problem Sets, templates—in an easily stored and navigated volume.

Practice

Each *Eureka Math* lesson begins with a series of energetic, joyous fluency activities, including those found in *Eureka Math Practice*. Students who are fluent in their math facts can master more material more deeply. With *Practice*, students build competence in newly acquired skills and reinforce previous learning in preparation for the next lesson.

Together, *Learn* and *Practice* provide all the print materials students will use for their core math instruction.

Succeed

Eureka Math Succeed enables students to work individually toward mastery. These additional problem sets align lesson by lesson with classroom instruction, making them ideal for use as homework or extra practice. Each problem set is accompanied by a Homework Helper, a set of worked examples that illustrate how to solve similar problems.

Teachers and tutors can use *Succeed* books from prior grade levels as curriculum-consistent tools for filling gaps in foundational knowledge. Students will thrive and progress more quickly as familiar models facilitate connections to their current grade-level content.

Students, families, and educators:

Thank you for being part of the *Eureka Math*™ community, where we celebrate the joy, wonder, and thrill of mathematics.

In the *Eureka Math* classroom, new learning is activated through rich experiences and dialogue. The *Learn* book puts in each student's hands the prompts and problem sequences they need to express and consolidate their learning in class.

What is in the Learn book?

Application Problems: Problem solving in a real-world context is a daily part of *Eureka Math*. Students build confidence and perseverance as they apply their knowledge in new and varied situations. The curriculum encourages students to use the RDW process—Read the problem, Draw to make sense of the problem, and Write an equation and a solution. Teachers facilitate as students share their work and explain their solution strategies to one another.

Problem Sets: A carefully sequenced Problem Set provides an in-class opportunity for independent work, with multiple entry points for differentiation. Teachers can use the Preparation and Customization process to select "Must Do" problems for each student. Some students will complete more problems than others; what is important is that all students have a 10-minute period to immediately exercise what they've learned, with light support from their teacher.

Students bring the Problem Set with them to the culminating point of each lesson: the Student Debrief. Here, students reflect with their peers and their teacher, articulating and consolidating what they wondered, noticed, and learned that day.

Exit Tickets: Students show their teacher what they know through their work on the daily Exit Ticket. This check for understanding provides the teacher with valuable real-time evidence of the efficacy of that day's instruction, giving critical insight into where to focus next.

Templates: From time to time, the Application Problem, Problem Set, or other classroom activity requires that students have their own copy of a picture, reusable model, or data set. Each of these templates is provided with the first lesson that requires it.

Where can I learn more about Eureka Math resources?

The Great Minds® team is committed to supporting students, families, and educators with an ever-growing library of resources, available at eureka-math.org. The website also offers inspiring stories of success in the *Eureka Math* community. Share your insights and accomplishments with fellow users by becoming a *Eureka Math* Champion.

Best wishes for a year filled with aha moments!

Jill Diniz

Jill Diniz
Director of Mathematics
Great Minds

The Read–Draw–Write Process

The *Eureka Math* curriculum supports students as they problem-solve by using a simple, repeatable process introduced by the teacher. The Read–Draw–Write (RDW) process calls for students to

1. Read the problem.
2. Draw and label.
3. Write an equation.
4. Write a word sentence (statement).

Educators are encouraged to scaffold the process by interjecting questions such as

- What do you see?
- Can you draw something?
- What conclusions can you make from your drawing?

The more students participate in reasoning through problems with this systematic, open approach, the more they internalize the thought process and apply it instinctively for years to come.

Contents

Module 1: Sums and Differences to 10

Read

Dora found 5 leaves that blew in through the window. Then, she found 2 more leaves that blew in. Draw a picture and use numbers to show how many leaves Dora found in all.

Draw

Lesson 1:   Analyze and describe embedded numbers (to 10) using 5-groups and number bonds.

©2018 Great Minds®. eureka-math.org

1

Write

Thear was
5 Pek suppose
and came 5 may prpole.

5 + 5 = 5

EUREKA MATH™

Name _____ Date _____

Circle 5, and then make a number bond.

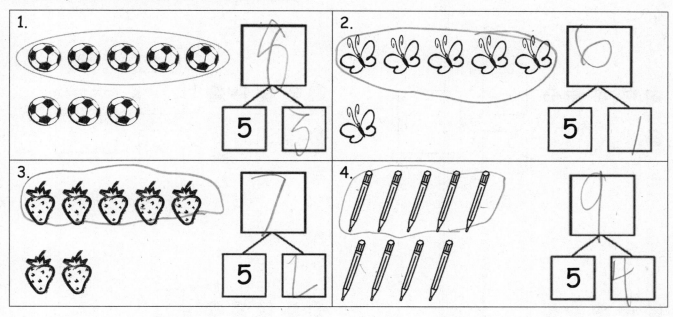

Put nail polish on the number of fingernails shown from left to right. Then, fill in the parts. Make the number of fingernails on one hand a part.

5.

8

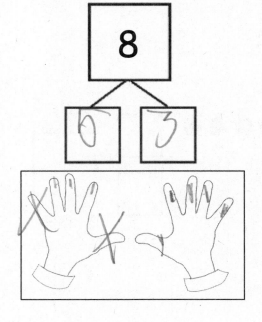

6.

6

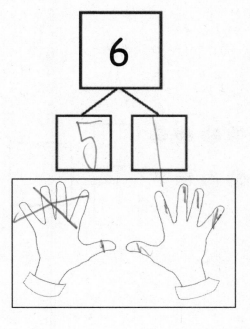

EUREKA MATH

Lesson 1: Analyze and describe embedded numbers (to 10) using 5-groups and number bonds.

©2018 Great Minds®. eureka-math.org

3

Make a number bond that shows 5 as one part.

7.

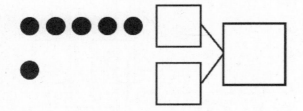

8.

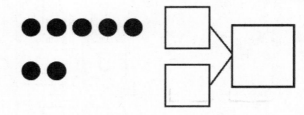

9.

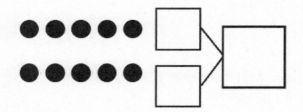

10.

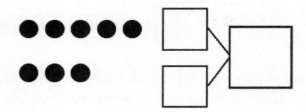

11.

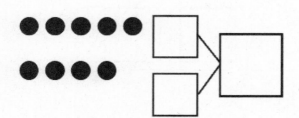

12.

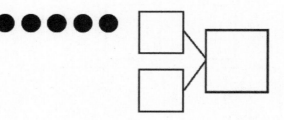

Lesson 1:   Analyze and describe embedded numbers (to 10) using 5-groups and number bonds.

EUREKA MATH™

Name _____   Date _____

Make a number bond for the pictures that shows 5 as one part.

1.

2.

EUREKA MATH™

Lesson 1:   Analyze and describe embedded numbers (to 10) using 5-groups and number bonds.

©2018 Great Minds®. eureka-math.org

5

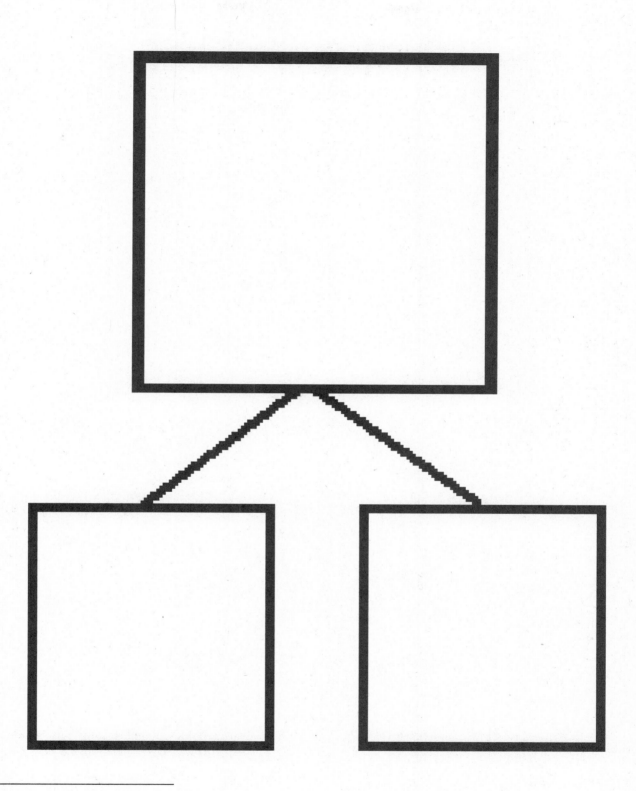

number bond

EUREKA MATH™

Lesson 1: Analyze and describe embedded numbers (to 10) using 5-groups and number bonds.

©2018 Great Minds®. eureka-math.org

7

Read

Bella spilled some pencils on the carpet. Geno came over to help her pick them up. Geno found 5 pencils under the desk and Bella found 4 by the door. How many pencils did they find together?

Draw a math picture and write a number bond and a number sentence that tells about the story.

Draw

Lesson 2:   Reason about embedded numbers in varied configurations using number bonds.

©2018 Great Minds®. eureka-math.org

9

Write

They found ▢ pencils.

Lesson 2:   Reason about embedded numbers in varied configurations using number bonds.

©2018 Great Minds®. eureka-math.org

EUREKA
MATH™

Name _____   Date _____

Circle 2 parts you see. Make a number bond to match.

1.

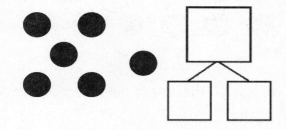

2.

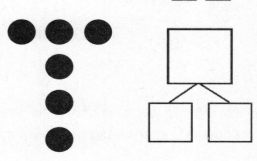

3.

4.

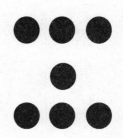

5.

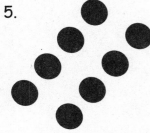

6.

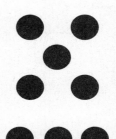
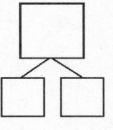

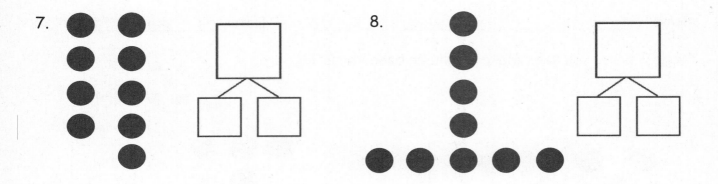

7.

8.

9. How many pieces of fruit do you see? Write at least 2 different number bonds to show different ways to break apart the total.

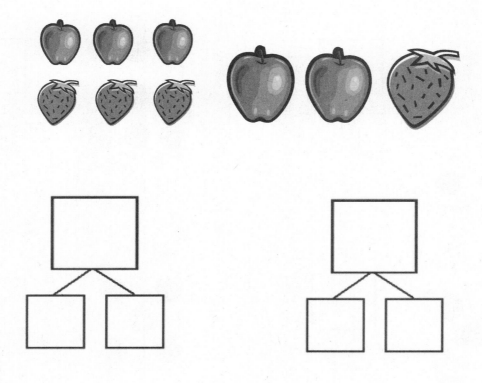

Lesson 2:   Reason about embedded numbers in varied configurations using number bonds.

©2018 Great Minds®. eureka-math.org

EUREKA MATH™

Name _____   Date _____

Circle 2 parts you see. Make a number bond to match.

1.

   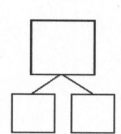

2.

   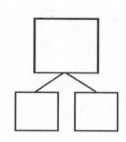

3.

   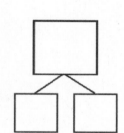

4.

   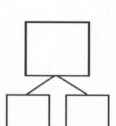

Read

Alex had 9 marbles in his hand. He hid his hands behind his back and put some in one hand and some in the other. How many marbles might be in each hand?

Use pictures or numbers to draw a number bond to show your idea.

Draw

EUREKA MATH™   **Lesson 3:**   See and describe numbers of objects using *1 more* within 5-group   **15**
                                                      configurations.

©2018 Great Minds®. eureka-math.org

Write

_____

_____

_____

Lesson 3: See and describe numbers of objects using *1 more* within 5-group configurations.

Name _____   Date _____

Draw one more in the 5-group. In the box, write the numbers to describe the new picture.

1.

1 more than 7 is _____.
7 + 1 = _____

2.

1 more than 9 is _____.
9 + 1 = _____

3.

1 more than 6 is _____.
6 + 1 = _____

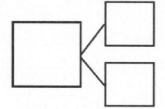

4.

1 more than 5 is _____.
5 + 1 = _____

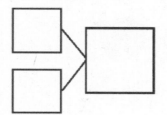

EUREKA MATH

Lesson 3:   See and describe numbers of objects using *1 more* within 5-group configurations.

17

©2018 Great Minds®. eureka-math.org

5.

1 more than 8 is _____.

8 + 1 = _____

6.

_____ is 1 more than 7.

_____ = 7 + 1

7.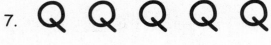

_____ is 1 more than 6.

_____ = 6 + 1

8.

_____ is 1 more than 5.

_____ = 5 + 1

9. Imagine adding 1 more backpack to the picture. Then, write the numbers to match how many backpacks there will be.

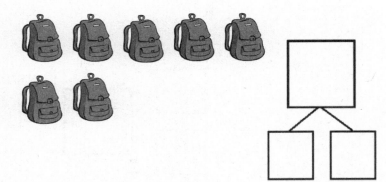

1 more than 7 is _____.

_____ + 1 = _____

Lesson 3:   See and describe numbers of objects using *1 more* within 5-group configurations.

EUREKA MATH™

Name _____   Date _____

How many objects do you see? Draw one more. How many objects are there now?

1.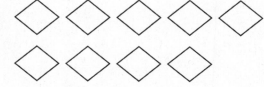

2.

```
_____ is 1 more than 9.

9 + 1 = _____
```

```
1 more than 6 is _____.

_____ + 1 = _____
```

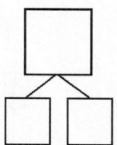

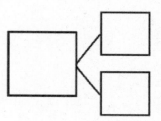

EUREKA MATH™

Lesson 3:   See and describe numbers of objects using *1 more* within 5-group configurations.

19

©2018 Great Minds®. eureka-math.org

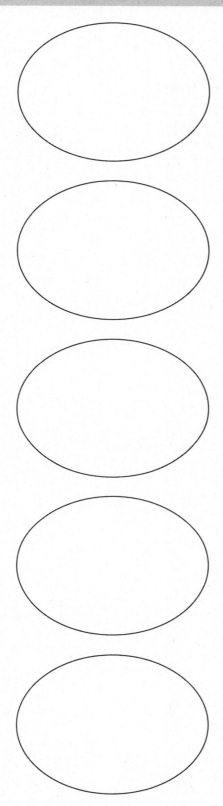
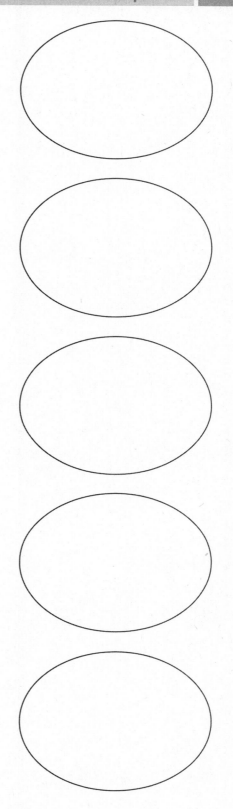

_____

5-group mat

Lesson 3:   See and describe numbers of objects using *1 more* within 5-group
                  configurations.

©2018 Great Minds®. eureka-math.org

21

Read

Our class had 4 pumpkins. On Monday, Marta brought 1 more pumpkin.

How many pumpkins did our class have on Monday?

On Tuesday, Beto brought 1 more pumpkin. How many pumpkins did our

class have on Tuesday?

Then, on Wednesday, Shea brought 1 more pumpkin. How many pumpkins

did our class have on Wednesday?

Draw a picture and write a number sentence to show your thinking. What

do you notice about what happened each day?

Extension: If this pattern continues, how many pumpkins will our class

have on Friday?

Lesson 4:   Represent *put together* situations with number bonds. Count on from
one embedded number or part to totals of 6 and 7, and generate all
addition expressions for each total.

©2018 Great Minds®. eureka-math.org

23

Draw

Write

_____

_____

_____

Lesson 4: Represent *put together* situations with number bonds. Count on from one embedded number or part to totals of 6 and 7, and generate all addition expressions for each total.

EUREKA MATH™

Name _____   Date _____

Ways to Make 6.

Use the apple picture to help you write all of the different ways to make 6.

☐ + ☐

☐ + ☐

☐ ⟨ ☐ ☐

☐ ☐ ⟩ ☐

☐ + ☐

☐ + ☐

☐ + ☐

☐ + ☐

☐ ⟨ ☐ ☐

☐ ☐ ⟩ ☐

☐ + ☐

☐ + ☐

EUREKA MATH

Lesson 4:   Represent *put together* situations with number bonds. Count on from one embedded number or part to totals of 6 and 7, and generate all addition expressions for each total.

©2018 Great Minds®. eureka-math.org

25

Name _____ Date _____

Show different ways to make 6. In each set, shade some circles and leave the others blank.

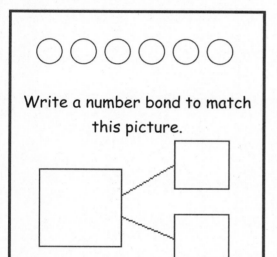

Write a number bond to match this picture.

Write a number sentence to match this picture.

☐ + ☐ = ☐

EUREKA MATH™   **Lesson 4:** Represent *put together* situations with number bonds. Count on from one embedded number or part to totals of 6 and 7, and generate all addition expressions for each total.

©2018 Great Minds®. eureka-math.org

27

6 apples picture card

Lesson 4:   Represent *put together* situations with number bonds. Count on from
one embedded number or part to totals of 6 and 7, and generate all
addition expressions for each total.

©2018 Great Minds®. eureka-math.org

29

Read

Marcus had 6 pieces of candy. He decided to give some to his mother and keep some for himself.

Use pictures and numbers to show two ways that Marcus could have split up 6 pieces of his candy.

Draw

Lesson 5: Represent *put together* situations with number bonds. Count on from one embedded number or part to totals of 6 and 7, and generate all addition expressions for each total.

©2018 Great Minds®. eureka-math.org

31

Write

_____

_____

_____

Lesson 5:   Represent *put together* situations with number bonds. Count on from one embedded number or part to totals of 6 and 7, and generate all addition expressions for each total.

EUREKA
MATH™

Name _____   Date _____

Ways to Make 7.   Use the classroom picture to help you write the expressions and number bonds to show all of the different ways to make 7.

Lesson 5:

Represent *put together* situations with number bonds. Count on from one embedded number or part to totals of 6 and 7, and generate all addition expressions for each total.

©2018 Great Minds®. eureka-math.org

33

EUREKA
MATH™

Name _____   Date _____

Color in two dice that make 7 together. Then, fill in the number bond and number sentences to match the dice you colored.

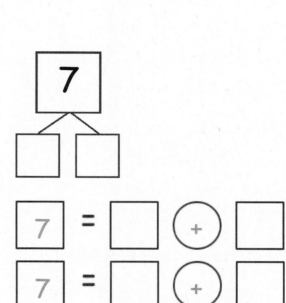

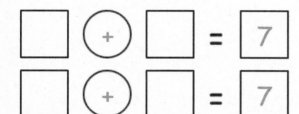

$$\boxed{} \; \bigoplus \; \boxed{} \; = \; \boxed{7}$$

$$\boxed{} \; \bigoplus \; \boxed{} \; = \; \boxed{7}$$

$$\boxed{7} \; = \; \boxed{} \; \bigoplus \; \boxed{}$$

$$\boxed{7} \; = \; \boxed{} \; \bigoplus \; \boxed{}$$

EUREKA MATH™

Lesson 5:   Represent *put together* situations with number bonds. Count on from one embedded number or part to totals of 6 and 7, and generate all addition expressions for each total.

35

7 children picture card

Lesson 5: Represent *put together* situations with number bonds. Count on from one embedded number or part to totals of 6 and 7, and generate all addition expressions for each total.

©2018 Great Minds®. eureka-math.org

37

Read

Tom has 4 red cars and 3 green cars. Dave has 5 red cars and 2 green cars. Dave thinks he has more cars than Tom has. Is Dave right? Draw a picture to show how you know. Write a number bond to show each of the boys' sets of cars.

Draw

Lesson 6:   Represent *put together* situations with number bonds. Count on from one embedded number or part to totals of 8 and 9, and generate all expressions for each total.

©2018 Great Minds®. eureka-math.org

39

Write

_____

_____

_____

Lesson 6: Represent *put together* situations with number bonds. Count on from one embedded number or part to totals of 8 and 9, and generate all expressions for each total.

EUREKA
MATH™

Name _____   Date _____

Circle the part. Count on to show 8 with the picture and number bond. Write the expressions.

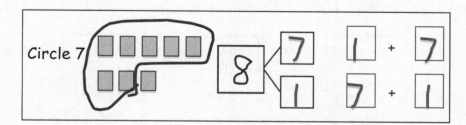

1. Circle 6. How many more does 6 need to make 8?

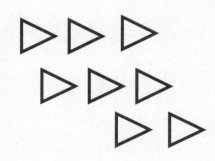

2. Circle 5. How many more does 5 need to make 8?

3. Circle 4. How many more does 4 need to make 8?

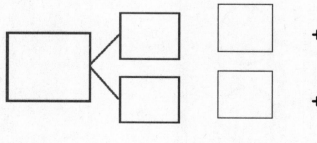

EUREKA MATH

Lesson 6:   Represent *put together* situations with number bonds. Count on from one embedded number or part to totals of 8 and 9, and generate all expressions for each total.

41

©2018 Great Minds®. eureka-math.org

4. These number bonds are in an order starting with the biggest part first. Write to show which number bonds are missing.

a. 8 / 8 0   b. 8 / 7 ☐   c. 8 / 6 ☐   d. 8 / ☐ 3   e. 8 / ☐ ☐

5. Use the expression to write a number bond and draw a picture that makes 8.

3 + 5

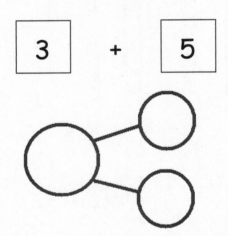

6. Use the expression to write a number bond and draw a picture that makes 8.

8 + 0

Lesson 6:   Represent *put together* situations with number bonds. Count on from one embedded number or part to totals of 8 and 9, and generate all expressions for each total.

EUREKA MATH

©2018 Great Minds®. eureka-math.org

Name _____ Date _____

Fill in the missing part of the number bond, and count on to find the total. Then, write 2 addition sentences for each number bond.

1.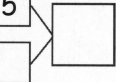

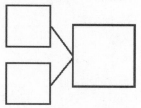

2.

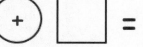

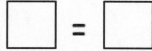

EUREKA MATH™

Lesson 6: Represent *put together* situations with number bonds. Count on from one embedded number or part to totals of 8 and 9, and generate all expressions for each total.

43

©2018 Great Minds®. eureka-math.org

8 animals picture card

EUREKA MATH™

Lesson 6:   Represent *put together* situations with number bonds. Count on from
one embedded number or part to totals of 8 and 9, and generate all
expressions for each total.

©2018 Great Minds®. eureka-math.org

45

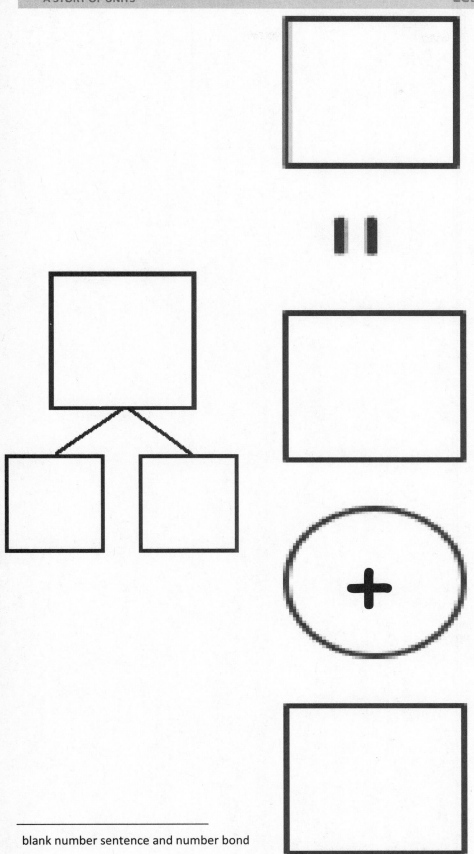

blank number sentence and number bond

Lesson 6:   Represent *put together* situations with number bonds. Count on from one embedded number or part to totals of 8 and 9, and generate all expressions for each total.

©2018 Great Minds®. eureka-math.org

47

Name _____   Date _____

Use your 5-group cards to help you write the expressions and number bonds to show all of the different ways to make 8.

☐ + ☐

☐ + ☐

☐ ⟨ ☐ ☐

☐ ☐ ⟩ ☐

☐ + ☐

☐ + ☐

☐ + ☐

☐ + ☐

☐ ⟨ ☐ ☐

☐ ☐ ⟩ ☐

☐ + ☐

☐ + ☐

☐ + ☐

☐ + ☐

☐ ⟨ ☐ ☐

_____
ways to make 8

Lesson 6:   Represent *put together* situations with number bonds. Count on from one embedded number or part to totals of 8 and 9, and generate all expressions for each total.

49

©2018 Great Minds®. eureka-math.org

Read

Jenny has 8 flowers in a vase. The flowers come in two different colors.
Draw a picture to show what the vase of flowers might look like. Write a
number sentence and a number bond to match your picture.

Draw

Lesson 7:   Represent *put together* situations with number bonds. Count on from
one embedded number or part to totals of 8 and 9, and generate all
expressions for each total.   51

©2018 Great Minds®. eureka-math.org

Write

Lesson 7:   Represent *put together* situations with number bonds. Count on from
one embedded number or part to totals of 8 and 9, and generate all
expressions for each total.

©2018 Great Minds®. eureka-math.org

EUREKA
MATH™

Name _____   Date _____

Circle the part. Count on to show 9 with the picture and number bond. Write the expressions.

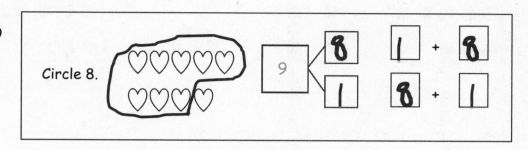

Circle 8.

1. Circle 7. How many more does 7 need to make 9?

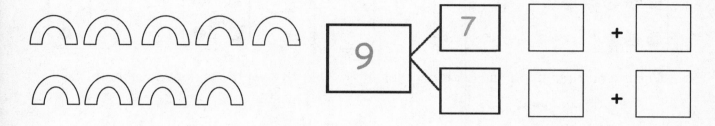

2. Circle 4. How many more does 4 need to make 9?

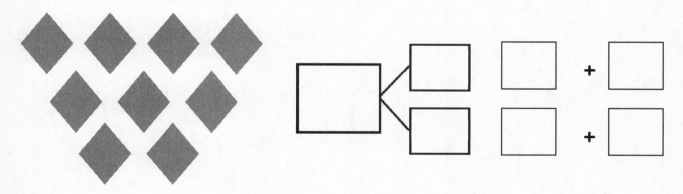

3. Circle 3. How many more does 3 need to make 9?

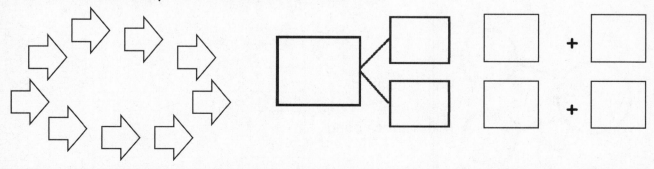

EUREKA MATH   Lesson 7:   Represent *put together* situations with number bonds. Count on from   53
                                      one embedded number or part to totals of 8 and 9, and generate all
                                      expressions for each total.

©2018 Great Minds®. eureka-math.org

4. Draw a line to show partners of 9.

a.    b.   c.   d.    e.

5. Write a number bond for each partner of 9. Use the partners above for help.

a. 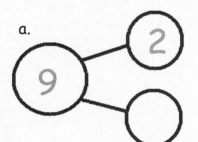   b.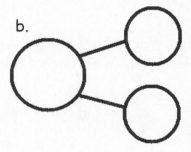

c. 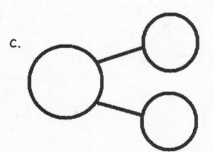   d.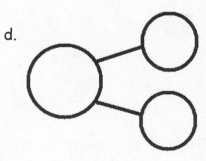

e. 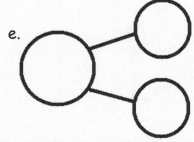 Write number sentences to match this number bond!

 + =

 + =

Lesson 7: Represent *put together* situations with number bonds. Count on from one embedded number or part to totals of 8 and 9, and generate all expressions for each total.

EUREKA MATH™

Name _____   Date _____

1. Circle the pairs of numbers that make 9.

2. Complete the number bonds to show 2 different ways to make 9.

a.

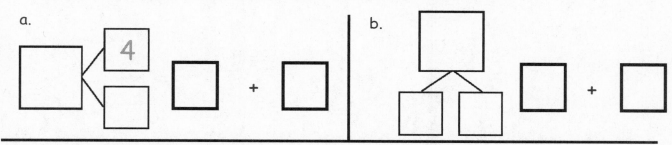

EUREKA
MATH™

Lesson 7:   Represent *put together* situations with number bonds. Count on from
one embedded number or part to totals of 8 and 9, and generate all
expressions for each total.

©2018 Great Minds®. eureka-math.org

55

9 books picture card

Lesson 7:   Represent *put together* situations with number bonds. Count on from
one embedded number or part to totals of 8 and 9, and generate all
expressions for each total.

©2018 Great Minds®. eureka-math.org

57

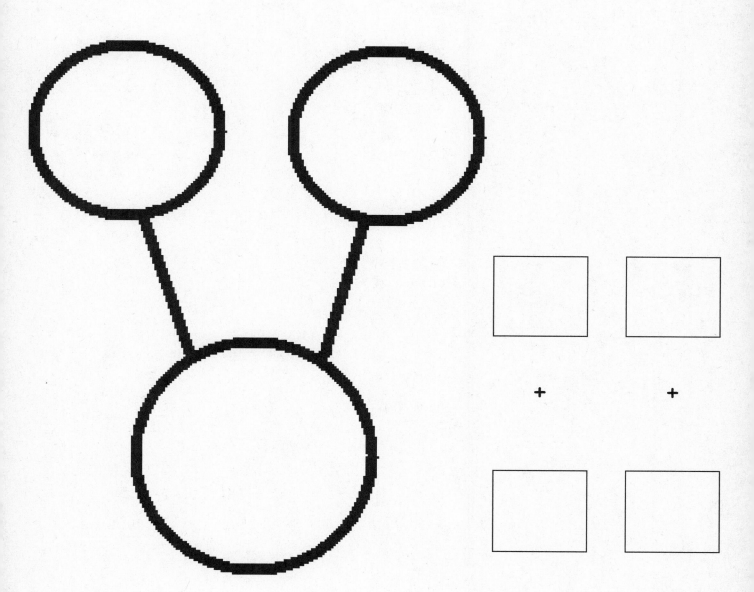

number bond and expression

Lesson 7: Represent *put together* situations with number bonds. Count on from
one embedded number or part to totals of 8 and 9, and generate all
expressions for each total.   59

©2018 Great Minds®. eureka-math.org

Read

Rayden received 9 stickers at school. He received 5 stickers in the morning. How many stickers did he receive in the afternoon?
Draw a picture, a number bond, and a number sentence to show how you know.

Draw

Lesson 8:   Represent all the number pairs of 10 as number bonds from a given
scenario, and generate all expressions equal to 10.

61

©2018 Great Minds®. eureka-math.org

Write

Rayden received stickers in the afternoon.

Lesson 8: Represent all the number pairs of 10 as number bonds from a given scenario, and generate all expressions equal to 10.

©2018 Great Minds®. eureka-math.org

EUREKA
MATH™

Name _____   Date _____

1. Use your bracelet to show different partners of 10. Then, draw the beads.
   Write an expression to match.

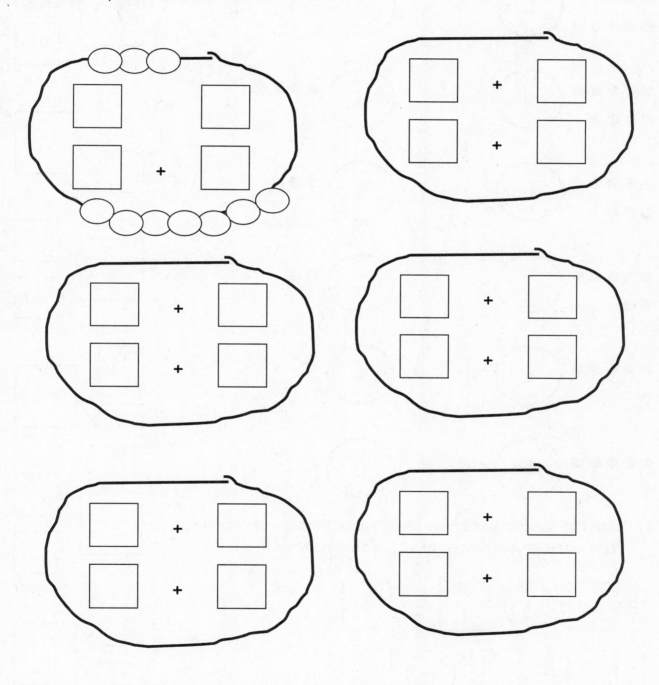

EUREKA
MATH™

Lesson 8:   Represent all the number pairs of 10 as number bonds from a given
scenario, and generate all expressions equal to 10.

©2018 Great Minds®. eureka-math.org

2. Match the partners of 10. Then, write a number bond for each partner.

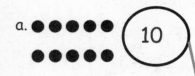 a. 10   5    a.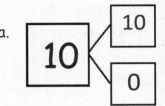

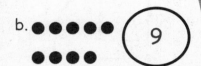 b. 9   4    b.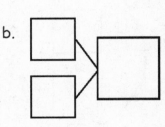

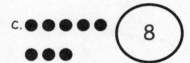 c. 8   3    c.

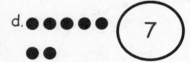 d. 7   2 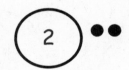   d.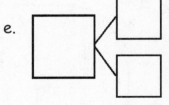

 e. 6   1    e.

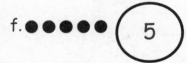 f. 5   0   f.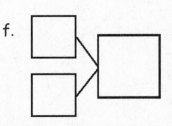

3. Color the number bond that has 2 parts that are the same.
Write addition sentences to match that number bond.

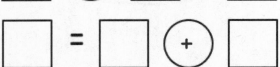

Lesson 8:   Represent all the number pairs of 10 as number bonds from a given
scenario, and generate all expressions equal to 10.

©2018 Great Minds®. eureka-math.org

EUREKA MATH™

Name _____   Date _____

Color the partners that make 10.

7 •••

•••• 6

8 ••

6 •••

1 9

5 4

EUREKA MATH

Lesson 8:   Represent all the number pairs of 10 as number bonds from a given
scenario, and generate all expressions equal to 10.

©2018 Great Minds®. eureka-math.org

65

Read

Kira was making a number bracelet with a total of 10 beads on it. She has put on 3 red beads so far. How many more beads does she need to add to the bracelet?

Explain your thinking in a picture and number sentence.

Draw

Lesson 9:   Solve *add to with result unknown* and *put together with result unknown* math stories by drawing, writing equations, and making statements of the solution.

©2018 Great Minds®. eureka-math.org

67

Write

Kira needs 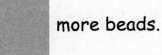 more beads.

Lesson 9:   Solve *add to with result unknown* and *put together with result unknown* math stories by drawing, writing equations, and making statements of the solution.
©2018 Great Minds®. eureka-math.org

EUREKA
MATH™

Name _____   Date _____

1.

[ ]   +   [ ]   =   [ ]

_____ balls are here.   _____ more roll over.   Now, there are _____ balls.

Make a number bond to match the story.

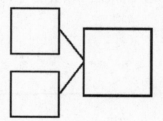

2.

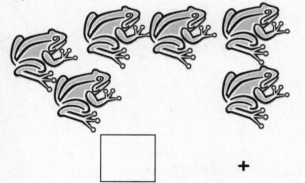

[ ]   +   [ ]   =   [ ]

_____ frogs are here.   _____ more hops over.   Now, there are _____ frogs.

Make a number bond to match the story.

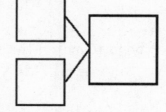

EUREKA MATH   Lesson 9:   Solve *add to with result unknown* and *put together with result unknown* math stories by drawing, writing equations, and making statements of the solution.   69

©2018 Great Minds®. eureka-math.org

3.

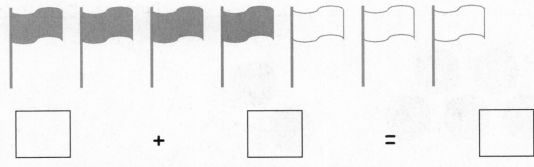

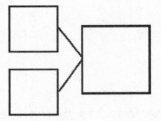

There are _____ dark flags.   There are ____ white flags.

Altogether, there are _____ flags.

Make a number bond to match the story.

4.

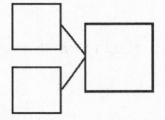

There are _____ white flowers.   There are ____ dark flowers.

Altogether, there are _____ flowers.

Make a number bond to match the story.

Lesson 9:   Solve *add to with result unknown* and *put together with result unknown* math stories by drawing, writing equations, and making statements of the solution.
©2018 Great Minds®. eureka-math.org

EUREKA MATH™

Name _____   Date _____

Draw a picture and write a number sentence to match the story.

Ben has 3 red balls and gets 5 green balls. How many balls does he have now?

```
┌──────────────────────────────────────────────────┐
│                                                    │
│                                                    │
│                                                    │
│                                                    │
│                                                    │
│                                                    │
└──────────────────────────────────────────────────┘
```

┌──────┐   ┌──────┐   ┌──────┐
│   │   +   │   │   =   │   │   Ben has _____ balls.
└──────┘   └──────┘   └──────┘

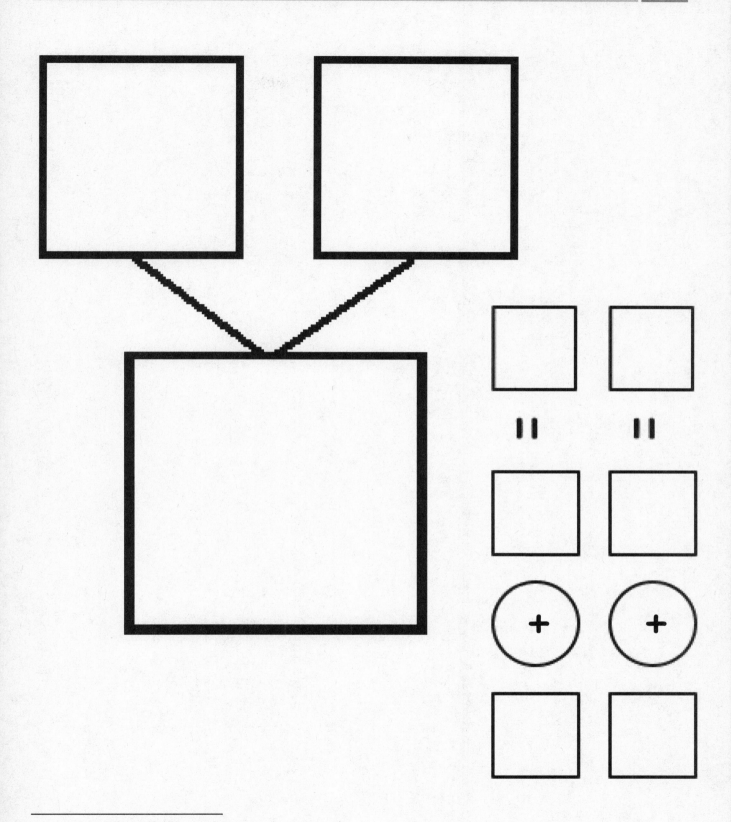

number bond and two blank equations

EUREKA
MATH™

Lesson 9:   Solve *add to with result unknown* and *put together with result unknown* math stories by drawing, writing equations, and making statements of the solution.

©2018 Great Minds®. eureka-math.org

73

Read

The class is collecting canned food to help those in need. The teacher brings in 3 cans to start the collection. On Monday, Becky brings in 2 cans. On Tuesday, Talia brings in 2 cans. On Wednesday, Brendan brings in 2 cans. How many cans were there at the end of each day?

Draw a picture to show your thinking. What do you notice about what happened each day?

Extension: If this pattern continues, how many cans will the class have on Friday?

    Lesson 10:   Solve *put together with result unknown* math stories by drawing and
using 5-group cards.

©2018 Great Minds®. eureka-math.org

75

Draw

Write

_____

_____

_____

Lesson 10:   Solve _put together with result unknown_ math stories by drawing and
                            using 5-group cards.

Name _____ Date _____

1. Use the picture to write the number sentence and the number bond.

 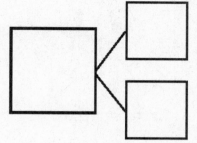

_____ little turtles + _____ big turtles = _____ turtles

2.

_____ dogs that are awake + _____ sleeping dogs = _____ dogs

3.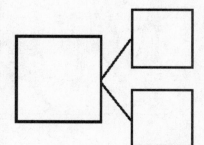

_____ pigs not in mud + _____ pigs in mud = _____ pigs

EUREKA MATH™

Lesson 10: Solve *put together with result unknown* math stories by drawing and using 5-group cards.

©2018 Great Minds®. eureka-math.org

77

4. Draw a line from the picture to the matching 5-group cards.

a.

b.

c.

d.

Lesson 10:   Solve *put together with result unknown* math stories by drawing and
                        using 5-group cards.

EUREKA
MATH

Name _____   Date _____

1. Draw to show the story. There are 3 large balls and 4 small balls.

How many balls are there? There are _____ balls.

2. Circle the set of tiles that match your picture.

EUREKA MATH   **Lesson 10:**   Solve *put together with result unknown* math stories by drawing and   79
                                        using 5-group cards.

                        ©2018 Great Minds®. eureka-math.org

Read

There are 8 children in the afterschool cooking club. How many boys and how many girls might be in the class? Draw a picture and write a number sentence to explain your thinking.

Extension: How many other combinations of boys and girls could be made? Write a number bond for each combination you can think of.

Draw

Lesson 11:   Solve *add to with change unknown* math stories as a context for counting on by drawing, writing equations, and making statements of the solution.

©2018 Great Minds®. eureka-math.org

81

Write

Lesson 11:   Solve *add to with change unknown* math stories as a context for counting on by drawing, writing equations, and making statements of the solution.

©2018 Great Minds®. eureka-math.org

EUREKA MATH™

Name _____ Date _____

1. Jill was given a total of 5 flowers for her birthday. Draw more flowers in the vase to show Jill's birthday flowers.

How many flowers did you have to draw? ____ flowers

Write a number sentence and a number bond to match the story.

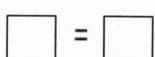  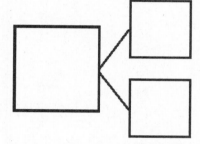

2. Kate and Nana were baking cookies. They made 2 heart cookies and then made some square cookies. They made 8 cookies altogether. How many square cookies did they make? Draw and count on to show the story.

Write a number sentence and a number bond to match the story.

  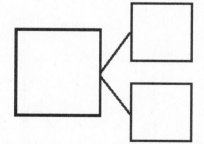

$2 \; \boxed{+} \; \square \; = \; 8$

EUREKA MATH

Lesson 11:   Solve *add to with change unknown* math stories as a context for counting on by drawing, writing equations, and making statements of the solution.

©2018 Great Minds®. eureka-math.org

83

Show the parts. Write a number bond to match the story.

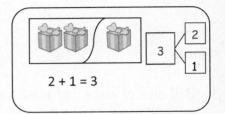

2 + 1 = 3

3. Bill has 2 trucks. His friend, James, came over with some more. Together, they had 5 trucks. How many trucks did James bring over?

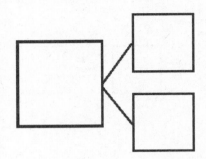

James brought over _____ trucks.

Write a number sentence to explain the story.

$$\boxed{2} \enspace \oplus \enspace \boxed{} \enspace = \enspace \boxed{5}$$

4. Jane caught 7 fish before she stopped to eat lunch. After lunch, she caught some more. At the end of the day, she had 9 fish. How many fish did she catch after lunch?

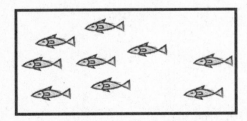
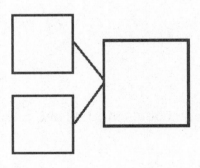

Jane caught _____ fish after lunch.

Write a number sentence to explain the story.

$$\boxed{} \enspace \oplus \enspace \boxed{} \enspace = \enspace \boxed{}$$

Lesson 11:   Solve *add to with change unknown* math stories as a context for counting on by drawing, writing equations, and making statements of the solution.
©2018 Great Minds®. eureka-math.org

Name _____ Date _____

Draw more bears to show that Jen has 8 bears total.

I added _____ more bears.

Write a number sentence to show how many
bears you drew.

 $+$ $=$

EUREKA MATH

Lesson 11: Solve *add to with change unknown* math stories as a context for
counting on by drawing, writing equations, and making statements of
the solution.

©2018 Great Minds®. eureka-math.org

85

Read

Tanya has 7 books on her shelf. She borrowed some books from the library, and now there are 9 books on her shelf. How many books did she get at the library?

Explain your thinking in pictures, words, or with a number sentence. Draw a box around the mystery number in your number sentence.

Draw

Lesson 12:   Solve *add to with change unknown* math stories using 5-group cards.

©2018 Great Minds®. eureka-math.org

87

Write

Tanya got [ ] books at the library.

Lesson 12:   Solve *add to with change unknown* math stories using 5-group cards.

Name _____   Date _____

Use your

4 •••••

5-group cards

Fill in the missing numbers.

1.

3   +   _____   =   5

2.

5   +   _____   =   9

3.

4   +   _____   =   10

4. Kate and Bob had 6 balls at the park. Kate had 2 of the balls.

   How many balls did Bob have?

   _____ balls   **=**   _____ balls   **+**   _____ balls

                    Bob had _____ balls at the park.

---

5. I had 3 apples. My mom gave me some more. Then, I had 10 apples.

   How many apples did my mom give me?

   _____ apples   **+**   _____ apples   **=**   _____ apples

   Mom gave me _____ apples.

Lesson 12:   Solve *add to with change unknown* math stories using 5-group cards.

Name _____   Date _____

Draw a picture, and count on to solve the math story.

Bob caught 5 fish. John caught some more fish. They had 7 fish in all. How many fish did John catch?

Write a number sentence to match your picture.

☐ **+** ☐ **=** ☐

John caught _____ fish.

Read

Sammi had 6 bunnies. One of them had babies. Now, she has 10 bunnies. How many babies were born?

Draw a picture to show how you know. Write a number bond and a number sentence to match your picture.

Draw

Lesson 13:   Tell *put together with result unknown, add to with result unknown, and add to with change unknown* stories from equations.

©2018 Great Minds®. eureka-math.org

93

Write

There were baby bunnies born.

Lesson 13:   Tell *put together with result unknown, add to with result unknown, and add to with change unknown* stories from equations.

©2018 Great Minds®. eureka-math.org

EUREKA MATH™

Name _____ Date _____

With a partner, create a story for each of the number sentences below. Draw a picture to show. Write the number bond to match the story.

1.   6 + 2 = ☐

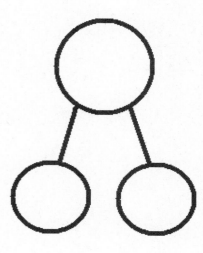

2.   5 + 5 = ☐

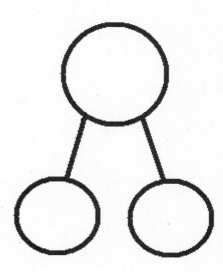

EUREKA MATH

Lesson 13: Tell *put together with result unknown, add to with result unknown, and add to with change unknown* stories from equations.

©2018 Great Minds®. eureka-math.org

95

3.   5 + ☐ = 7

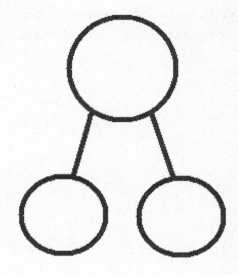

---

4.   6 + ☐ = 10

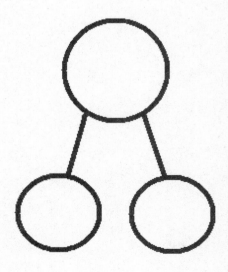

Lesson 13:   Tell *put together with result unknown, add to with result unknown, and add to with change unknown* stories from equations.

©2018 Great Minds®. eureka-math.org

EUREKA
MATH™

Name _____   Date _____

Tell a math story for each number sentence by drawing a picture.

1. $5 + 1 = 6$

2. $3 + ? = 8$

EUREKA MATH™   **Lesson 13:**   Tell *put together with result unknown, add to with result unknown,*   **97**
and add to with change unknown stories from equations.

©2018 Great Minds®. eureka-math.org

Read

Beth went apple picking. She picked 7 apples and put them in her basket.
Two more apples fell out of the tree right into her basket! How many
apples does she have in her basket now?
Draw a math picture and write a number bond and number sentence to
match the story.

Draw

Lesson 14:   Count on up to 3 more using numeral and 5-group cards and fingers
to track the change.

©2018 Great Minds®. eureka-math.org

99

Write

Beth has [ ] apples in her basket.

Lesson 14:   Count on up to 3 more using numeral and 5-group cards and fingers to track the change.

Name _____   Date _____

1. Count on to add.

$\boxed{}$ $\oplus$ $\boxed{}$ = $\boxed{}$ There are _____ flowers altogether.

2.

$\boxed{}$ = $\boxed{}$ $\oplus$ $\boxed{}$ There are _____ oranges in all.

3.

$\boxed{}$ = $\boxed{}$ $\oplus$ $\boxed{}$ There is a total of _____ crayons.

EUREKA MATH

Lesson 14:   Count on up to 3 more using numeral and 5-group cards and fingers
to track the change.

©2018 Great Minds®. eureka-math.org

101

4. Use your 5-group cards to count on to add. Try to use as few dot cards as you can.

a. $6 + 1 = \boxed{}$

b. $6 + 3 = \boxed{}$

c. $7 + 2 = \boxed{}$

d. $\boxed{} = 5 + 3$

5. Use your 5-group cards, your fingers, or your known facts to count on to add.

a. $8 + 2 = \boxed{}$

b. $\boxed{} = 4 + 1$

c. $4 + 3 = \boxed{}$

d. $\boxed{} = 6 + 3$

Lesson 14:   Count on up to 3 more using numeral and 5-group cards and fingers
to track the change.

©2018 Great Minds®. eureka-math.org

EUREKA MATH

Name _____   Date _____

1.

6

$6 + 2 = \boxed{}$

I counted _____ hats in all.

---

2. Count on to solve the number sentences.

a.

$7 + 3 = \boxed{}$

b.

$8 + 2 = \boxed{}$

EUREKA MATH™   **Lesson 14:**   Count on up to 3 more using numeral and 5-group cards and fingers   **103**
                                       to track the change.

©2018 Great Minds®. eureka-math.org

Read

Joshua and Rebecca were eating raisins. Joshua had 7 raisins and took 2 more from the box. Rebecca had 9 raisins and took 2 more from the box. Who had a greater number of raisins, Joshua or Rebecca?

Draw math drawings and write number bonds or number sentences to show how you know.

Draw

Lesson 15:   Count on up to 3 more using numeral and 5-group cards and fingers
to track the change.

©2018 Great Minds®. eureka-math.org

105

Write

_____

_____

_____

Lesson 15:   Count on up to 3 more using numeral and 5-group cards and fingers
to track the change.

©2018 Great Minds®. eureka-math.org

Name _____   Date _____

1. Count on to add.

a.

   ☐ ⊕ ☐ = ☐   There are ____ crayons altogether.

b.

☐ ⊕ ☐ = ☐   There are a total of ____ balloons.

c.

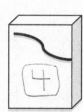   

☐ = ☐ ⊕ ☐   In all, there are ____ pencils.

EUREKA MATH   **Lesson 15:**   Count on up to 3 more using numeral and 5-group cards and fingers   107
                                 to track the change.

©2018 Great Minds®. eureka-math.org

2. What shortcut or efficient strategy can you find to add?

a. 4 (+) 1 = ☐   h. 2 (+) 5 = ☐

b. 4 (+) 3 = ☐   i. 7 (+) 2 = ☐

c. 7 (+) 1 = ☐   j. 7 (+) 3 = ☐

d. ☐ = 6 (+) 2   k. ☐ = 4 (+) 2

e. ☐ = 5 (+) 3   l. ☐ = 2 (+) 5

f. ☐ = 3 (+) 6   m. ☐ = 6 (+) 2

g. ☐ = 3 (+) 7   n. ☐ = 2 (+) 8

Lesson 15:   Count on up to 3 more using numeral and 5-group cards and fingers
to track the change.

©2018 Great Minds®. eureka-math.org

EUREKA
MATH™

Name _____   Date _____

Use the picture to add.

6

☐ + ☐ = ☐

Show the shortcut you used to add.

There are _____ eggs total.

EUREKA MATH

Lesson 15:   Count on up to 3 more using numeral and 5-group cards and fingers to track the change.

109

©2018 Great Minds®. eureka-math.org

Read

There were 10 bowling pins standing. Finn knocked over some bowling pins, and 7 were still standing. How many did he knock over?

Use a simple math drawing to show what you did to solve. Write a number sentence with a box to show the mystery or unknown number.

Draw

Lesson 16:   Count on to find the unknown part in missing addend equations such as 6 + ___ = 9. Answer, "How many more to make 6, 7, 8, 9, and 10?"   **111**

©2018 Great Minds®. eureka-math.org

Write

_____

_____

_____

Count on to find the unknown part in missing addend equations such as 6 + ___ = 9. Answer, "How many more to make 6, 7, 8, 9, and 10?"

EUREKA
MATH™

Name _____   Date _____

1. Draw more apples to solve 4 + ? = 6.

I added _____ apples to the tree.

2. How many more to make 7?

3. How many more to make 8?

4. How many more to make 9?

©2018 Great Minds®. eureka-math.org

$$\boxed{3} \;\oplus\; \boxed{1} \;=\; \boxed{4}$$

5. Count on to add. (Circle) the strategy you used to keep track.

a. $\boxed{4} \;\oplus\; \boxed{} \;=\; \boxed{5}$

b. $\boxed{4} \;\oplus\; \boxed{} \;=\; \boxed{7}$

c. $\boxed{8} \;=\; \boxed{5} \;\oplus\; \boxed{}$

d. $\boxed{10} \;=\; \boxed{} \;\oplus\; \boxed{8}$

e. $\boxed{7} \;\oplus\; \boxed{} \;=\; \boxed{8}$

f. $\boxed{} \;\oplus\; \boxed{5} \;=\; \boxed{7}$

g. $\boxed{8} \;=\; \boxed{6} \;\oplus\; \boxed{}$

h. $\boxed{10} \;=\; \boxed{} \;\oplus\; \boxed{7}$

Lesson 16:   Count on to find the unknown part in missing addend equations such as 6 + __ = 9. Answer, "How many more to make 6, 7, 8, 9, and 10?"

EUREKA MATH™

Name _____   Date _____

Solve the number sentences. (Circle) the tool or strategy you used.

a. $5 + \boxed{} = \boxed{7}$

I counted on using

Or

I just knew

---

b. $6 + \boxed{} = \boxed{9}$

I counted on using

Or

I just knew

EUREKA MATH™

Lesson 16:   Count on to find the unknown part in missing addend equations such
as 6 + ___ = 9. Answer, "How many more to make 6, 7, 8, 9, and 10?"

115

©2018 Great Minds®. eureka-math.org

Read

There are 10 swings on the playground, and 7 students are using the swings. How many swings are empty?

Draw or write a number sentence to show your thinking. Use a sentence at the end to answer today's question: How many swings are empty?

Draw

Lesson 17:   Understand the meaning of the equal sign by pairing equivalent
expressions and constructing true number sentences.

117

©2018 Great Minds®. eureka-math.org

Write

_____

_____

_____

Lesson 17: Understand the meaning of the equal sign by pairing equivalent expressions and constructing true number sentences.

Name _____   Date _____

Write an expression that matches the groups on each plate. If the plates have the same amount of fruit, write the equal sign between the expressions.

[ ] + [ ]   (=)   [ ] + [ ]
2   3   1   4

1.

[ ] + [ ]   ( )   [ ] + [ ]

2.

[ ] + [ ]   ( )   [ ] + [ ]

3.

[ ] + [ ]   ( )   [ ] + [ ]

4.

[ ] + [ ]   ( )   [ ] + [ ]

EUREKA MATH

Lesson 17:   Understand the meaning of the equal sign by pairing equivalent expressions and constructing true number sentences.

119

©2018 Great Minds®. eureka-math.org

5. Write an expression to match each domino.

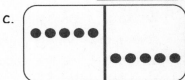
2+5

a.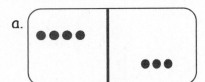

b.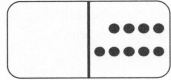

c.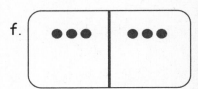

_____   _____   _____

d.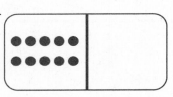

e.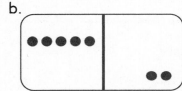

f.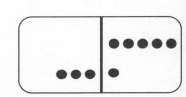

_____   _____   _____

g. Find two sets of expressions from (a)–(f) that are equal. Connect them below with = to make true number sentences.

_____   _____

6. a.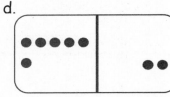

b.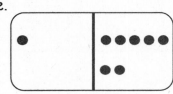

c.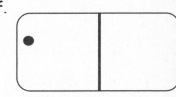

_____   _____   _____

d.

e.

f.

_____   _____   _____

g. Find two sets of expressions from (a)–(f) that are equal. Connect them below with = to make true number sentences.

_____   _____

Lesson 17:   Understand the meaning of the equal sign by pairing equivalent
expressions and constructing true number sentences.

EUREKA
MATH™

Name _____ Date _____

1. Use math drawings to make the pictures equal. Connect them below with = to make true number sentences.

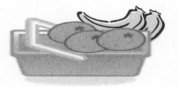   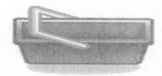

_____   _____

2. Shade the equal dominoes. Write a true number sentence.

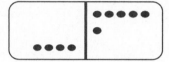   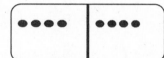   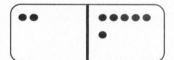

_____   _____

Lesson 17:   Understand the meaning of the equal sign by pairing equivalent expressions and constructing true number sentences.

©2018 Great Minds®. eureka-math.org

121

EUREKA
MATH™

Read

Dylan has 4 cats and 2 dogs at home. Laura has 1 dog and 5 fish at home.
Laura says she and Dylan have an equal number of pets. Dylan thinks he
has more pets than Laura. Who is right?

Draw a picture, write two number bonds, and use a number sentence to
show if Dylan and Laura have an equal amount of pets.

Draw

Lesson 18:   Understand the meaning of the equal sign by pairing equivalent
expressions and constructing true number sentences.

123

©2018 Great Minds®. eureka-math.org

Write

Lesson 18: Understand the meaning of the equal sign by pairing equivalent expressions and constructing true number sentences.

©2018 Great Minds®. eureka-math.org

EUREKA MATH

Name _____ Date _____

1. Add. Color the balloons that match the number in the boy's mind. Find expressions that are equal. Connect them below with = to make true number sentences.

a.

_____

b.

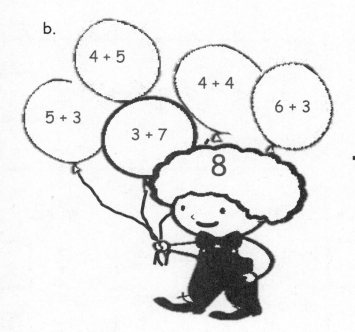

_____

EUREKA MATH

Lesson 18:   Understand the meaning of the equal sign by pairing equivalent
             expressions and constructing true number sentences.

©2018 Great Minds®. eureka-math.org

125

2. Are these number sentences true? if it is true. if it is false.

If it is false, rewrite the number sentence to make it true.

a. 3 + 1 = 2 + 2 ☐

b. 9 + 1 = 1 + 2

_____

_____

c. 2 + 3 = 1 + 4 ☐

d. 5 + 1 = 4 + 2

_____

_____

e. 4 + 3 = 3 + 5 ☐

f. 0 + 10 = 2 + 8

_____

_____

g. 6 + 3 = 4 + 5 ☐

h. 3 + 7 = 2 + 6

_____

_____

3. Write a number in the expression and solve. if it is true. if it is false.

a. 1 + ___ = 3 + 2 ☐

b. ___ + 4 = 2 + 5

c. ___ + 5 = 6 + ___ ☐

d. 7 + ___ = 8 + ___

Lesson 18: Understand the meaning of the equal sign by pairing equivalent expressions and constructing true number sentences.

©2018 Great Minds®. eureka-math.org

EUREKA MATH™

Name _____   Date _____

Find two ways to fix each number sentence to make it true.

a.

| 7 + 3 = 6 + 2 |

b.

| 8 + 1 = 3 + 5 |

7 + 3 = 6 + 4

_____ _____

_____ _____

_____ _____

_____ _____

Lesson 18:   Understand the meaning of the equal sign by pairing equivalent
expressions and constructing true number sentences.

127

©2018 Great Minds®. eureka-math.org

Read

Dylan has 4 cats and 2 dogs at home. Sammy has 1 mama bunny and 6 baby bunnies at home.

Draw a number bond showing the total number of pets of each household.

Write a statement to tell if the two households have an equal number of pets.

Draw

Lesson 19:   Represent the same story scenario with addends repositioned (the commutative property).

©2018 Great Minds®. eureka-math.org

129

Write

_____

_____

_____

Lesson 19:   Represent the same story scenario with addends repositioned (the commutative property).

©2018 Great Minds®. eureka-math.org

EUREKA
MATH™

Name _____ Date _____

1. Write the number bond to match the picture. Then, complete the number sentences.

a.

☐ (+) ☐ = 5   ☐ = ☐ (+) ☐

☐ (+) ☐ = 5   ☐ = ☐ (+) ☐

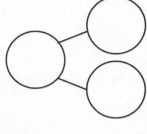

b.

☐ (+) ☐ = 8   8 = ☐ (+) ☐

☐ (+) ☐ = ☐   ☐ = ☐ (+) ☐

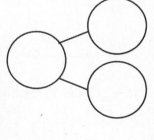

c.

☐ (+) ☐ = ☐   ☐ = ☐ (+) ☐

☐ (+) ☐ = ☐   ☐ = ☐ (+) ☐

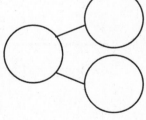

EUREKA MATH

Lesson 19:   Represent the same story scenario with addends repositioned (the commutative property).

©2018 Great Minds®. eureka-math.org

131

Write the expression under each plate. Add the equal sign to show they are the same amount.

2.

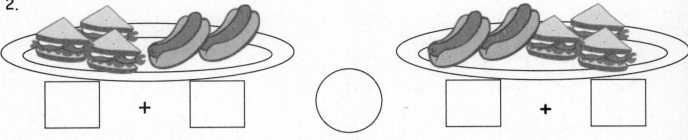

$\boxed{}$ + $\boxed{}$ $\bigcirc$ $\boxed{}$ + $\boxed{}$

3.

$\boxed{}$ + $\boxed{}$ $\bigcirc$ $\boxed{}$ + $\boxed{}$

4. Draw to show the expression.

  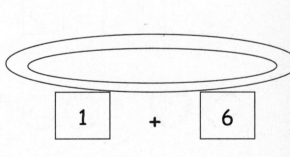

$\boxed{}$ + $\boxed{}$ $\bigcirc$ $\boxed{1}$ + $\boxed{6}$

5. Draw and write to show 2 expressions that use the same numbers and have the same tota

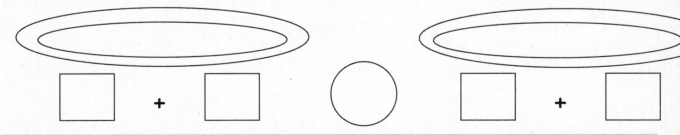

$\boxed{}$ + $\boxed{}$ $\bigcirc$ $\boxed{}$ + $\boxed{}$

Lesson 19:   Represent the same story scenario with addends repositioned (the commutative property).

Name _____   Date _____

Use the picture and write the number sentences to show the parts in a different order.

_____ + _____ = _____   _____ = _____ + _____

_____ + _____ = _____   _____ = _____ + _____

EUREKA
MATH™   Lesson 19:   Represent the same story scenario with addends repositioned (the   133
                              commutative property).

©2018 Great Minds®. eureka-math.org

Read

Laura had 5 fish. Her mother gave her 1 more. Laura's brother Frank had 1 fish. Their mother gave Frank 5 more. Laura cried, "That's not fair! He has more fish than I do!"

Use number bonds and a number sentence to show Laura the truth. If you can, write a sentence with words that would help Laura understand.

Draw

Write

_____

_____

_____

EUREKA MATH™

Name _____   Date _____

Circle the larger amount and count on. Write the number sentence, starting with the larger number.

1.

$2 + 7 = 9$

Color the larger part, and complete the number bond.
Write the number sentence, starting with the larger part.

2.

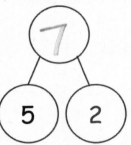

$2 + 5 = 7$

3.

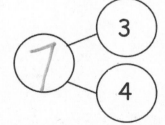

$3 + 4 = 7$

4.

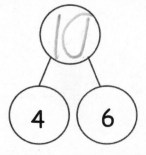

$6 + 4 = 10$

EUREKA MATH

Lesson 20:   Apply the commutative property to count on from a larger addend.

137

©2018 Great Minds®. eureka-math.org

Color the larger part of the bond. Count on from that part to find the total, and fill in the number bond. Complete the first number sentence, and then rewrite the number sentence to start with the larger part.

5.

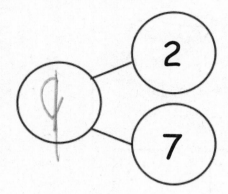

| 2 | $+$ | 5 | $=$ | 7 |

| 2 | $+$ | 7 | $=$ | 9 |

6.

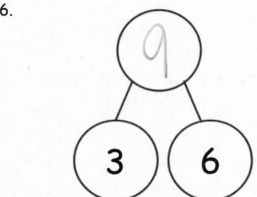

| 3 | $+$ | 6 | $=$ | 9 |

| 3 | $+$ | 8 | $=$ | 9 |

Circle the larger number, and count on to solve.

7. $1 + 5 =$ ___6___

8. $2 + 6 =$ ___9___

9. $4 + 3 =$ ___7___

10. $3 + 6 =$ ___9___

Lesson 20: Apply the commutative property to count on from a larger addend.

©2018 Great Minds®. eureka-math.org

EUREKA MATH™

Name _____ Date _____

Circle the larger part, and complete the number bond. Write the number sentence, starting with the larger part.

a.

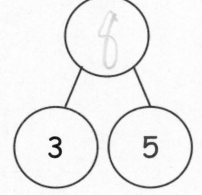

  + =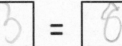

b.

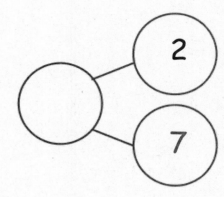

 + =

Lesson 20: Apply the commutative property to count on from a larger addend.

139

Read

Jordan is holding a container with 3 pencils. His teacher gives him 4 more pencils for the container. How many pencils will be in the container? Write a number bond, number sentence, and statement to show the solution.

Draw

EUREKA MATH

Lesson 21:   Visualize and solve doubles and doubles plus 1 with 5-group cards.

©2018 Great Minds®. eureka-math.org

141

Write

_____

_____

_____

Lesson 22:   Look for and make use of repeated reasoning on the addition chart by
solving and analyzing problems with common addends.

EUREKA
MATH™

Name _____   Date _____

Add the numbers on the pairs of cards. Write the number sentences. Color doubles red. Color doubles plus 1 blue.

1.

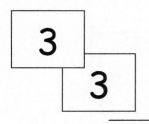

_____

2.

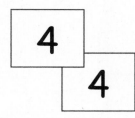

_____

3.

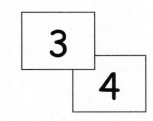

_____

4.

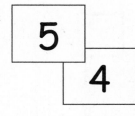

_____

Solve. Use your doubles to help. Draw and write the double that helped.

5. $5 + 4 = \boxed{}$ OOOOO
                                      OOOOO   _____

6. $4 + 3 = \boxed{}$ OOOOO
                                      OOOOO   _____

7. Solve the doubles and the doubles plus 1 number sentences.

a. $0 + 0 = \boxed{}$ $0 + 1 = \boxed{}$

b. $2 + 2 = \boxed{}$ $2 + 3 = \boxed{}$

c. $3 + 3 = \boxed{}$ $3 + 4 = \boxed{}$

d. $4 + 4 = \boxed{}$ $4 + 5 = \boxed{}$

e. $3 + \boxed{} = 6$ $3 + \boxed{} = 7$

f. $5 + \boxed{} = 10$ $4 + \boxed{} = 9$

8. Show how this strategy can help you solve $5 + 6 = \boxed{}$

9. Write a set of 4 related addition facts for the number sentences of Problem 7(d).

Lesson 21:   Visualize and solve doubles and doubles plus 1 with 5-group cards.

EUREKA MATH™

Name _____   Date _____

Write the double and double plus 1 number sentence for each 5-group card.

⋮	4	5

_____   _____   _____

_____   _____   _____

1+9									
1+8	2+8								
1+7	2+7	3+7							
1+6	2+6	3+6	4+6						
1+5	2+5	3+5	4+5	5+5					
1+4	2+4	3+4	4+4	5+4	6+4				
1+3	2+3	3+3	4+3	5+3	6+3	7+3			
1+2	2+2	3+2	4+2	5+2	6+2	7+2	8+2		
1+1	2+1	3+1	4+1	5+1	6+1	7+1	8+1	9+1	
1+0	2+0	3+0	4+0	5+0	6+0	7+0	8+0	9+0	10+0

addition chart

Lesson 21:   Visualize and solve doubles and doubles plus 1 with 5-group cards.

147

©2018 Great Minds®. eureka-math.org

Read

May and Kay are twins. Whatever May has, Kay has it, too. May has 2 dolls. How many dolls do May and Kay have together? May has 3 stuffed animals. How many stuffed animals do they have together?

Write a number bond, number sentence, and statement to show your solution.

Extension: If all the dolls and all the stuffed animals were put together for an imaginary tea party, how many toys would there be? Draw or write to explain your thinking.

Lesson 22:   Look for and make use of repeated reasoning on the addition chart by   149
solving and analyzing problems with common addends.

©2018 Great Minds®. eureka-math.org

Draw

Write

_____

_____

_____

Lesson 22:   Look for and make use of repeated reasoning on the addition chart by solving and analyzing problems with common addends.

Name _____   Date _____

1. Use RED to color boxes with 0 as an addend. Find the total for each.
2. Use ORANGE to color boxes with 1 as an addend. Find the total for each.
3. Use YELLOW to color boxes with 2 as an addend. Find the total for each.
4. Use GREEN to color boxes with 3 as an addend. Find the total for each.
5. Use BLUE to color the boxes that are left. Find the total for each.

orange

6 + 1

7

1 + 0	1 + 1	1 + 2	1 + 3	1 + 4	1 + 5	1 + 6	1 + 7	1 + 8	1 + 9
2 + 0	2 + 1	2 + 2	2 + 3	2 + 4	2 + 5	2 + 6	2 + 7	2 + 8	
3 + 0	3 + 1	3 + 2	3 + 3	3 + 4	3 + 5	3 + 6	3 + 7		
4 + 0	4 + 1	4 + 2	4 + 3	4 + 4	4 + 5	4 + 6			
5 + 0	5 + 1	5 + 2	5 + 3	5 + 4	5 + 5				
6 + 0	6 + 1	6 + 2	6 + 3	6 + 4					
7 + 0	7 + 1	7 + 2	7 + 3						
8 + 0	8 + 1	8 + 2							
9 + 0	9 + 1								
10 + 0									

Lesson 22:   Look for and make use of repeated reasoning on the addition chart by
               solving and analyzing problems with common addends.

151

Name _____ Date _____

Some of the addends in this chart are missing! Fill in the missing numbers.

1 + 0	1 + 1	1 + 2	1 + 3	1 + 4	1 + 5	1 + 6	1 + 7	1 + 8	1 + 9
2 + 0	2 + 1	2 + 2	2 + __	2 + 4	2 + 5	2 + 6	2 + 7	2 + 8	
3 + 0	3 + 1	3 + 2	3 + __	3 + 4	3 + 5	3 + 6	3 + 7		
4 + 0	4 + __	4 + 2	4 + 3	__ + 4	__ + 5	__ + 6			
5 + 0	5 + __	5 + 2	5 + 3	5 + 4	5 + 5				
6 + 0	6 + __	6 + 2	6 + 3	6 + 4					
7 + __	7 + 1	7 + 2	7 + 3						
8 + __	8 + 1	8 + 2							
9 + __	9 + 1								
10 + 0									

EUREKA MATH

Lesson 22: Look for and make use of repeated reasoning on the addition chart by solving and analyzing problems with common addends.

©2018 Great Minds®. eureka-math.org

153

Read

John has 3 stickers. Mark has 4 stickers. Anna has 5 stickers. They each get two more stickers. How many do they each have now?

Write a number bond and number sentence for each student.

Extension: How many stickers do John, Mark, and Anna have together?

Draw

Lesson 23: Look for and make use of structure on the addition chart by looking for and coloring problems with the same total.

155

©2018 Great Minds®. eureka-math.org

Write

_____

_____

_____

Look for and make use of structure on the addition chart by looking
for and coloring problems with the same total.

Name _____   Date _____

Use your chart to write a list of number sentences in the spaces below.

Totals of 10	Totals of 9	Totals of 8	Totals of 7

Lesson 23:   Look for and make use of structure on the addition chart by looking
for and coloring problems with the same total.

157

©2018 Great Minds®. eureka-math.org

Name _____   Date _____

1. Circle all the boxes that total 10.
2. Draw an X through all the boxes that total 8.

1 + 0	1 + 1	1 + 2	1 + 3	1 + 4	1 + 5	1 + 6	1 + 7	1 + 8	1 + 9
2 + 0	2 + 1	2 + 2	2 + 3	2 + 4	2 + 5	2 + 6	2 + 7	2 + 8	
3 + 0	3 + 1	3 + 2	3 + 3	3 + 4	3 + 5	3 + 6	3 + 7		
4 + 0	4 + 1	4 + 2	4 + 3	4 + 4	4 + 5	4 + 6			
5 + 0	5 + 1	5 + 2	5 + 3	5 + 4	5 + 5				
6 + 0	6 + 1	6 + 2	6 + 3	6 + 4					
7 + 0	7 + 1	7 + 2	7 + 3						
8 + 0	8 + 1	8 + 2							
9 + 0	9 + 1								
10 + 0									

EUREKA MATH™   **Lesson 23:**   Look for and make use of structure on the addition chart by looking   **159**
for and coloring problems with the same total.

©2018 Great Minds®. eureka-math.org

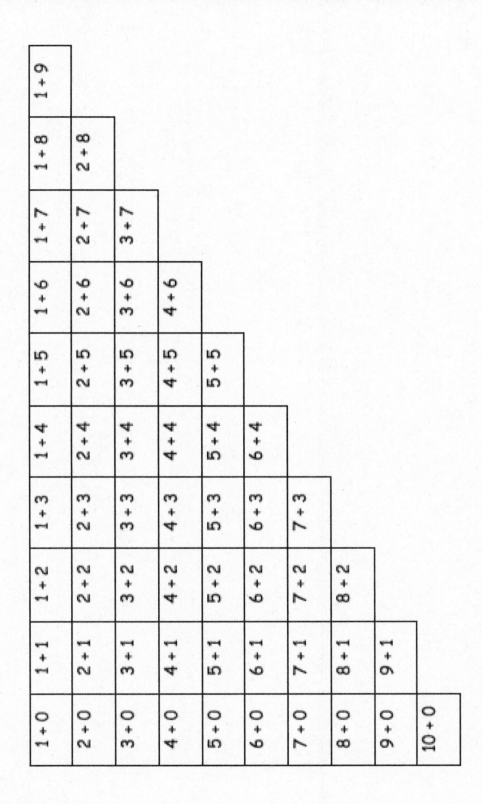

addition chart; from Lesson 21

Lesson 23:   Look for and make use of structure on the addition chart by looking
for and coloring problems with the same total.

161

©2018 Great Minds®. eureka-math.org

Read

The teacher told Henry to get 8 linking cubes. Henry took 4 blue cubes and 3 red cubes. Does Henry have the correct amount of linking cubes? Use pictures or words to explain your thinking.

Draw

Write

_____

_____

_____

Lesson 24:   Practice to build fluency with facts to 10.

EUREKA MATH

Name _____   Date _____

Related Fact Ladders

1.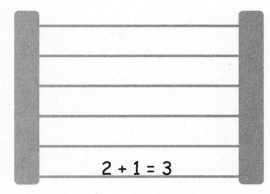

$2 + 1 = 3$

2.

$4 + 1 = 5$

3.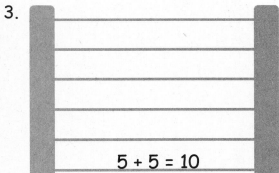

$5 + 5 = 10$

4.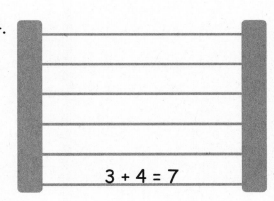

$3 + 4 = 7$

5.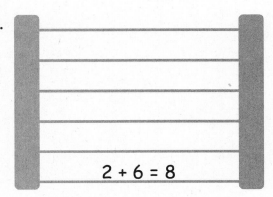

$2 + 6 = 8$

6.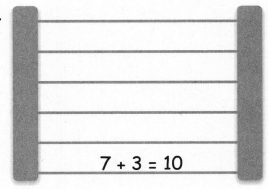

$7 + 3 = 10$

Name _____   Date _____

Solve the number sentences. Use the key to color. Once the box is colored, you do not need to color it again.

a. $5 + 2 =$ _____	b. $7 + 2 =$ _____	c. $2 + 3 =$ _____
d. $3 + 3 =$ _____	e. $7 = 1 +$ _____	f. $2 = 1 +$ _____
g. _____ $= 4 + 4$	h. $8 + 2 =$ _____	i. $3 + 4 =$ _____
j. _____ $= 5 + 4$	k. $10 = 1 +$ _____	l. $10 = 5 +$ ___

Color doubles red.

Color +1 blue.

Color +2 green.

Color doubles +1 brown.

Challenge:

List the number sentences that can be colored more than 1 way.

_____   _____

Read

Taylor and her sister Reilly each got 4 books from the library. Then, Reilly went back in and checked out another book. How many books do Taylor and Reilly have together?

Draw and label a number bond to show the part of the books Taylor took out and the part that Reilly took out. Write a statement to share your answer.

Lesson 25:   Solve *add to with change unknown* math stories with addition, and relate to subtraction. Model with materials, and write corresponding number sentences.

©2018 Great Minds®. eureka-math.org

169

Draw

Write

_____

_____

_____

Lesson 25:   Solve *add to with change unknown* math stories with addition, and relate to subtraction. Model with materials, and write corresponding number sentences.

©2018 Great Minds®. eureka-math.org

EUREKA MATH

Name _____   Date _____

Break the total into parts. Write a number bond and addition and subtraction number sentences to match the story.

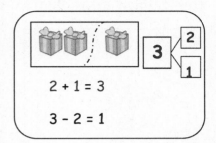

2 + 1 = 3

3 – 2 = 1

1. Rachel and Lucy are playing with 5 trucks. If Rachel is playing with 2 of them, how many is Lucy playing with?

   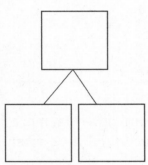

$2 \;\bigcirc{+}\; \square \;=\; 5$

$5 \;\bigcirc{-}\; 2 \;=\; \square$

Lucy is playing with _____ trucks.

2. Jane caught 9 fish. She caught 7 fish before she ate lunch. How many fish did she catch after lunch?

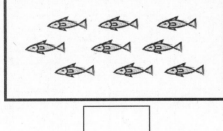

$\square \;\bigcirc{+}\; \square \;=\; 9$

$9 \;\bigcirc{-}\; \square \;=\; \square$

Jane caught _____ fish after lunch.

EUREKA MATH

Lesson 25:   Solve *add to with change unknown* math stories with addition, and relate to subtraction. Model with materials, and write corresponding number sentences.

©2018 Great Minds®. eureka-math.org

171

3. Dad bought 6 shirts. The next day he returned some of them. Now, he has 2 shirts. How many shirts did Dad return?

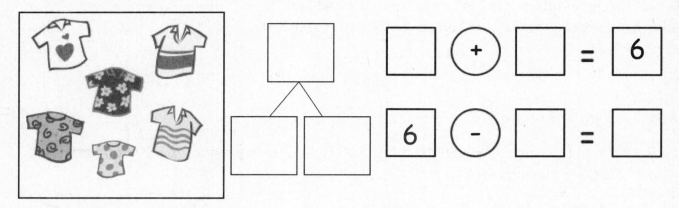

$$\boxed{} + \boxed{} = \boxed{6}$$

$$\boxed{6} - \boxed{} = \boxed{}$$

Dad returned _____ shirts.

4. John had 3 strawberries. Then, his friend gave him more fruit. Now, John has 7 pieces of fruit. How many pieces of fruit did John's friend give him?

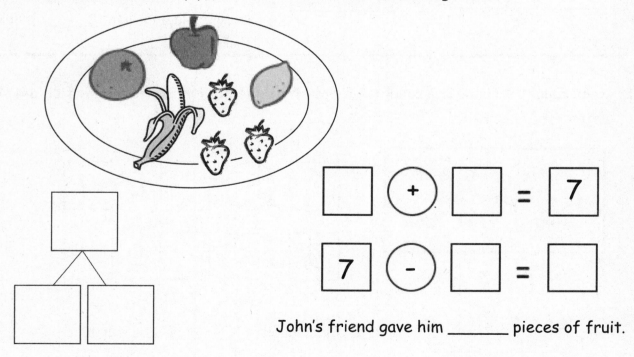

$$\boxed{} + \boxed{} = \boxed{7}$$

$$\boxed{7} - \boxed{} = \boxed{}$$

John's friend gave him _____ pieces of fruit.

Lesson 25:   Solve *add to with change unknown* math stories with addition, and relate to subtraction. Model with materials, and write corresponding number sentences.
©2018 Great Minds®. eureka-math.org

EUREKA MATH™

Name _____   Date _____

Solve the math story. Complete the number bond and number sentences. Color the unknown number yellow.

Rich bought 6 cans of soda on Monday.
He bought some more on Tuesday.
Now, he has 9 cans of soda.
How many cans did Rich buy on Tuesday?

Rich bought _____ cans.

EUREKA MATH™

Lesson 25: Solve *add to with change unknown* math stories with addition, and relate to subtraction. Model with materials, and write corresponding number sentences.

©2018 Great Minds®. eureka-math.org

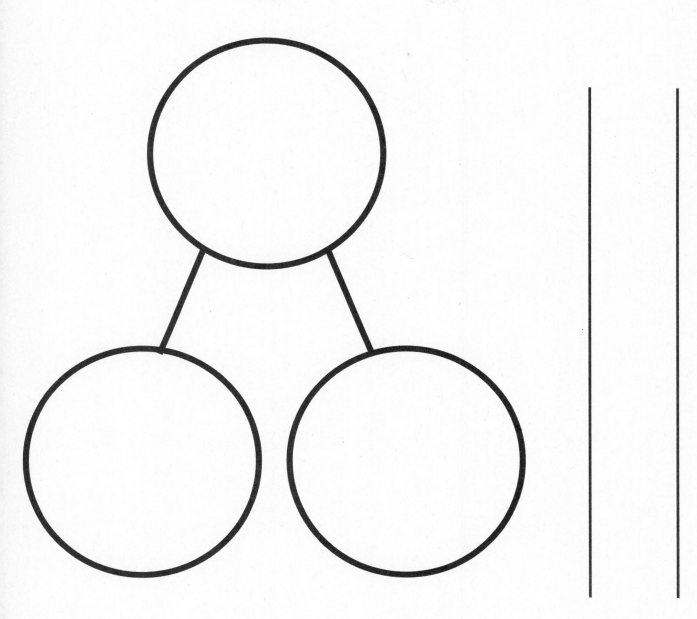

number bond and number sentences

EUREKA MATH

Lesson 25: Solve *add to with change unknown* math stories with addition, and relate to subtraction. Model with materials, and write corresponding number sentences.

©2018 Great Minds®. eureka-math.org

175

Read

There were 5 students in the cafeteria. Some more students came in late.

Now, there are 7 students in the cafeteria. How many students came

in late?

Write a number bond to match the story. Write an addition sentence and

a subtraction sentence to show two ways to solve the problem. Draw a

rectangle around the unknown number that you found.

Draw

Write

_____

_____

_____

Lesson 26:   Count on using the number path to find an unknown part.

Name _____   Date _____

Use the number path to solve.

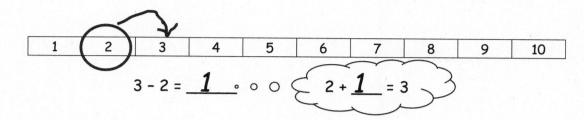

| 1 | 2 | 3 | 4 | 5 | 6 | 7 | 8 | 9 | 10 |

3 – 2 = __*1*__ ∘ ○ ○ (2 + __*1*__ = 3)

1.
| 1 | 2 | 3 | 4 | 5 | 6 | 7 | 8 | 9 | 10 |

6 – 4 = _____ ∘ ○ ○ (4 + _____ = 6)

2.
| 1 | 2 | 3 | 4 | 5 | 6 | 7 | 8 | 9 | 10 |

8 – 5 = _____ ∘ ○ ○ (5 + _____ = 8)

3.
| 1 | 2 | 3 | 4 | 5 | 6 | 7 | 8 | 9 | 10 |

9 – 6 = _____ ∘ ○ ○ (6 + _____ = 9)

4.
| 1 | 2 | 3 | 4 | 5 | 6 | 7 | 8 | 9 | 10 |

9 – 3 = _____ ∘ ○ ○ (3 + _____ = 9)

EUREKA
MATH™

Lesson 26:   Count on using the number path to find an unknown part.

179

Use the number path to help you solve.

1	2	3	4	5	6	7	8	9	10

5. $5 - 4 =$ _____ $4 +$ _____ $= 5$

6. $5 - 1 =$ _____ $1 +$ _____ $= 5$

7. $7 - 5 =$ _____ $5 +$ _____ $= 7$

8. $10 - 6 =$ _____ $6 +$ _____ $= 10$

9. $9 - 3 =$ _____ $3 +$ _____ $= 9$

Lesson 26:   Count on using the number path to find an unknown part.

EUREKA MATH

Name _____   Date _____

Use the number path to solve. Write the addition sentence you used to help you solve.

| 1 | 2 | 3 | 4 | 5 | 6 | 7 | 8 | 9 | 10 |

a. 7 – 5 = _____   _____

b. 9 – 2 = _____   _____

c. _____ = 10 – 3   _____

EUREKA MATH™

Lesson 26:   Count on using the number path to find an unknown part.

181

©2018 Great Minds®. eureka-math.org

| 1 | 2 | 3 | 4 | 5 | 6 | 7 | 8 | 9 | 10 |

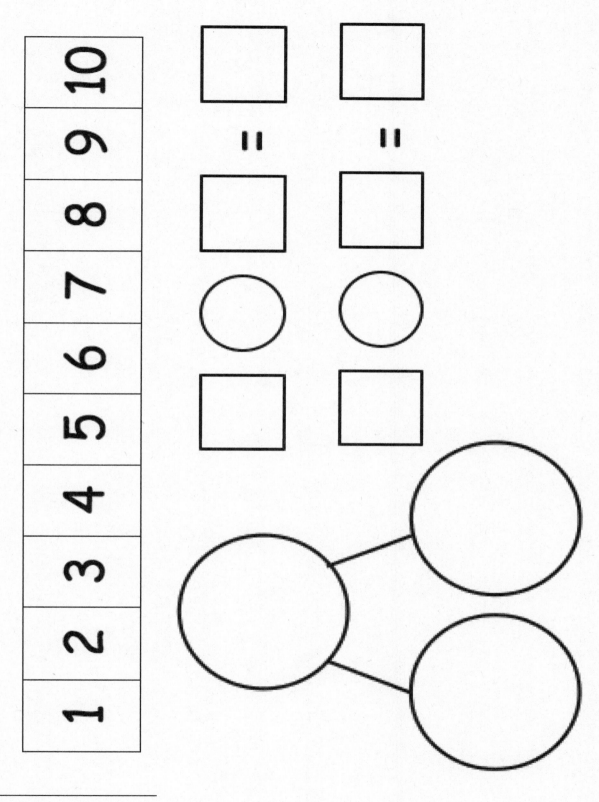

number path

Read

Marcus has 9 strawberries. Six of them are small; the rest are big. How many strawberries are big?

Fill in the template. Circle the mystery, or unknown, number in the number sentences, and write a statement to answer the question.

Draw

| 1 | 2 | 3 | 4 | 5 | 6 | 7 | 8 | 9 | 10 |

□ ○ □ = □

□ ○ □ = □

Write

_____

_____

_____

EUREKA MATH

Name _____   Date _____

| 1 | 2 | 3 | 4 | 5 | 6 | 7 | 8 | 9 | 10 |

Rewrite the subtraction number sentence as an addition number sentence.

Place a ☐ around the unknown. Use the number path if you want to.

1. $4 - 3 =$ ☐   _____ + _____ = _____

2. $6 - 2 =$ ☐   _____ + _____ = _____

3. $7 - 3 =$ ☐   _____ + _____ = _____

4. $9 - 6 =$ ☐   _____

5. $10 - 2 =$ ☐   _____

Use the number path to count on.

6. $8 - 4 =$ _____ $4 +$ _____ $= 8$

7. $9 - 5 =$ _____ $5 +$ _____ $= 9$

EUREKA MATH

Lesson 27:   Count on using the number path to find an unknown part.

187

©2018 Great Minds®. eureka-math.org

1	2	3	4	5	6	7	8	9	10

Hop back on the number path to count back.

8. 10 – 1 = _____

9. 9 – 2 = _____

10. Pick the best way to solve the problem. Check the box.

Count on   Count back

a. 10 – 9 = _____   ☐   ☐

b. 9 – 1 = _____   ☐   ☐

c. 8 – 5 = _____   ☐   ☐

d. 8 – 6 = _____   ☐   ☐

e. 7 – 4 = _____   ☐   ☐

f. 6 – 3 = _____   ☐   ☐

EUREKA MATH

Name _____   Date _____

To solve 7 - 6, Ben thinks you should count back, and Pat thinks you should count on. Which is the best way to solve this expression? Make a simple math drawing to show why.

$$7 - 6 = \text{\underline{\hspace{3cm}}}$$

Lesson 27:   Count on using the number path to find an unknown part.

189

©2018 Great Minds®. eureka-math.org

Read

Eight ducks are swimming in the pond. Four ducks fly away. How many ducks are still swimming in the pond?

Write a number bond, number sentence, and statement. Draw a number path to prove your answer.

Draw

Lesson 28:   Solve *take from with result unknown* math stories with math drawings, true number sentences, and statements, using horizontal marks to cross off what is taken away.

©2018 Great Minds®. eureka-math.org

191

Write

_____

_____

_____

Lesson 28:   Solve *take from with result unknown* math stories with math drawings, true number sentences, and statements, using horizontal marks to cross off what is taken away.

©2018 Great Minds®. eureka-math.org

EUREKA MATH™

Name _____   Date _____

Read the story. Draw a horizontal line through the items that are leaving the story.

Then, complete the number bond, sentence, and statement.

Example: 3 – 2 = 1

1. There are 5 toy airplanes flying at the park.
    One went down and broke.
    How many airplanes are still flying?

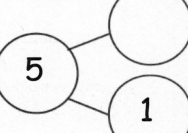

5 – 1 = _____

There are _____ airplanes still flying.

2. I had 6 eggs from the store.
    Three of them were cracked.
    How many eggs did I have that were not cracked?

6 – ____ = _____

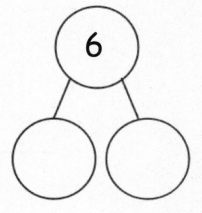

_____ eggs were not cracked.

EUREKA MATH

Lesson 28: Solve *take from with result unknown* math stories with math drawings, true number sentences, and statements, using horizontal marks to cross off what is taken away.

©2018 Great Minds®. eureka-math.org

193

Draw a number bond and math drawing to help you solve the problems.

3. Kate saw 8 cats playing in the grass.
    Three went away to chase a mouse.
    How many cats remained in the grass?

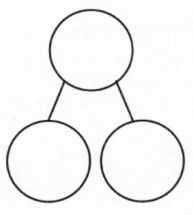

_____ - _____ = _____

_____ cats remained in the grass.

4. There were 7 mango slices.
    Two of them were eaten.
    How many mango slices are left to eat?

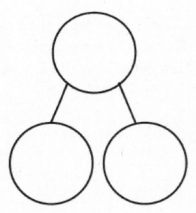

_____ - _____ = _____

There are _____ mango slices left.

   Lesson 28:   Solve *take from with result unknown* math stories with math drawings, true number sentences, and statements, using horizontal marks to cross off what is taken away.
©2018 Great Minds®. eureka-math.org

Name _____   Date _____

Read the problem. Make a math drawing to solve.

There were 9 kites flying in the park. Three kites got caught in trees. How many kites were still flying?

____ - ____ = ____

____ kites were still flying.

EUREKA MATH

Lesson 28:   Solve *take from with result unknown* math stories with math drawings, true number sentences, and statements, using horizontal marks to cross off what is taken away.

©2018 Great Minds®. eureka-math.org

Read

Lucas has 9 pencils for school. He lends 4 of them to his friends. How many pencils does Lucas have left?

Box the solution in your number sentence, and include a statement to answer the question. Be sure to draw your simple shapes in a straight line.

Draw

Write

_____

_____

_____

Lesson 29:   Solve *take apart with addend unknown* math stories with math drawings, equations, and statements, circling the known part to find the unknown.
©2018 Great Minds®. eureka-math.org

EUREKA
MATH™

Name _____   Date _____

Complete the story and solve. Label the number bond.
Color the missing part in the number sentence and number bond.

1. There are _____ apples.

    _____ have worms. Yuck!

    How many good apples are there?

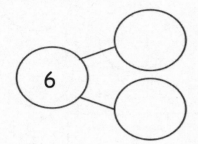

$$\boxed{6} - \boxed{} = \boxed{}$$

There are _____ good apples.

---

2. _____ books are in the case.

    _____ books are on the top shelf.

    How many books are on the bottom shelf?

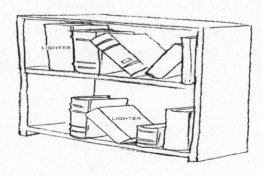

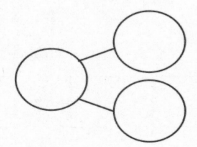

$$\boxed{9} - \boxed{} = \boxed{}$$

_____ books are on the bottom shelf.

EUREKA MATH

Lesson 29: Solve *take apart with addend unknown* math stories with math
drawings, equations, and statements, circling the known part to find
the unknown.

©2018 Great Minds®. eureka-math.org

199

Use number bonds and math drawings in a line to solve.

Example of math drawing and number sentence

$5 - 4 = 1$

3. There are 8 animals at the pond.
   Two are big. The rest are small.
   How many are small?

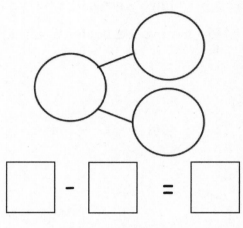

☐ − ☐ = ☐

_____ animals are small.

4. There are 7 students in the class.
   _____ students are girls.
   How many students are boys?

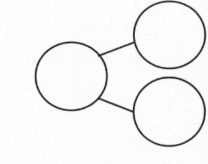

☐ − ☐ = ☐

_____ students are boys.

Lesson 29:   Solve *take apart with addend unknown* math stories with math
                    drawings, equations, and statements, circling the known part to find
                    the unknown.

EUREKA
MATH

Name _____   Date _____

Read the story. Make a math drawing to solve.

There are 9 baseball players on the team. Seven are on the bench. How many are not on the bench?

____ - ____ = ____

_____ players are not on the bench.

Read

Freddie has 10 action figures in his pocket. Five of them are good guys.

How many of his action figures are bad guys?

Box the solution in your number sentence, and include a statement to

answer the question. Make a math drawing. Circle the part that is good

guys to show you have the correct number of bad guys.

Draw

Lesson 30:   Solve *add to with change unknown* math stories with drawings,
                  relating addition and subtraction.

©2018 Great Minds®. eureka-math.org

203

Write

_____

_____

_____

    Lesson 30:   Solve *add to with change unknown* math stories with drawings, relating addition and subtraction.

EUREKA
MATH™

Name _____   Date _____

Solve the math stories. Complete and label the number bond and the picture number bond.   Lightly shade in the solution.

1. Jill was given a total of 5 flowers for her birthday. She put 3 in one vase and the rest in another vase. How many flowers did she put in the other vase?

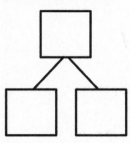

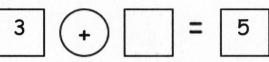

2. Kate and Nana were baking cookies. They made 5 heart-shaped cookies and then made some square cookies. They made 8 cookies altogether. How many square cookies did they make? Draw and solve.

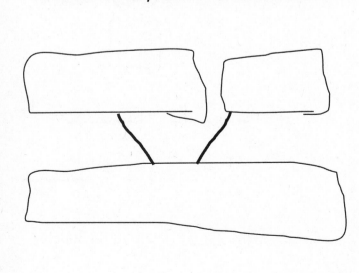

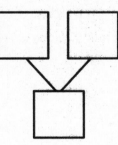

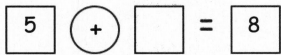

EUREKA MATH™   Lesson 30:   Solve *add to with change unknown* math stories with drawings, relating addition and subtraction.   205

©2018 Great Minds®. eureka-math.org

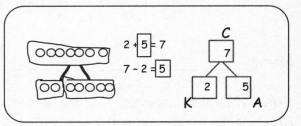

Solve. Complete and label the number bond and the picture number bond. Circle the unknown number.

3. Bill has 2 trucks. His friend James came over with some more.
Together, they have 6 trucks.
How many trucks did James bring over?

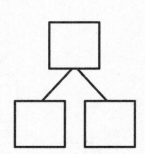

_____ + _____ = 6

6 - _____ = _____

James brought over _____ trucks.

___

4. Jane caught 5 fish before she stopped to eat lunch.
After lunch, she caught some more.
At the end of the day, she had 9 fish.
How many fish did she catch after lunch?

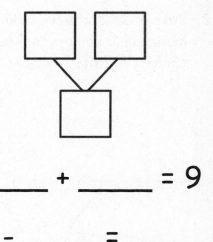

_____ + _____ = 9

9 - _____ = _____

Jane caught _____ fish after lunch.

Lesson 30: Solve *add to with change unknown* math stories with drawings, relating addition and subtraction.

©2018 Great Minds®. eureka-math.org

EUREKA MATH™

Name _____   Date _____

Draw and label a picture number bond to solve.

Toby collects shells. On Monday, he finds 6 shells. On Tuesday, he finds some more.
Toby finds a total of 9 shells. How many shells does Toby find on Tuesday?

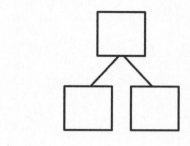

_____ + _____ = _____

_____ - _____ = _____

Toby finds _____ shells on Tuesday.

EUREKA MATH™   **Lesson 30:**   Solve *add to with change unknown* math stories with drawings,   **207**
                                              relating addition and subtraction.

©2018 Great Minds®. eureka-math.org

Read

Shanika saw 5 pigeons on the roof. Some more pigeons flew onto the roof.

She then counted 8 pigeons. How many pigeons flew over?

Write a number bond and both addition and subtraction number sentences

to match the story. Box the solution in your number sentences, and include

a statement to answer the question.

Draw

Write

Lesson 31: Solve *take from with change unknown* math stories with drawings.

EUREKA MATH™

Name _____   Date _____

Make a math drawing, and circle the part you know. Cross out the unknown part.

Complete the number sentence and number bond.

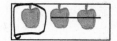

Sample: 3 – 1 = 2

1. Kate made 7 cookies. Bill ate some. Now, Kate has 5 cookies. How many cookies did Bill eat?

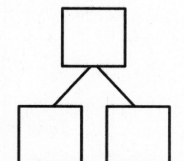

$7 \;\bigcirc\!\!-\; \square = \square$

Bill ate _____ cookies.

2. On Monday, Tim had 8 pencils. On Tuesday, he lost some pencils. On Wednesday, he has 4 pencils. How many pencils did Tim lose?

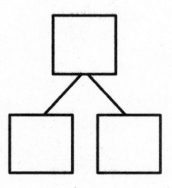

Tim lost _____ pencils.

$\square \;\bigcirc\!\!-\; \square = \square$

EUREKA MATH

Lesson 31:   Solve *take from with change unknown* math stories with drawings.

211

©2018 Great Minds®. eureka-math.org

3. A store had 6 shirts on the rack. Now, there are 2 shirts on the rack.
   How many shirts were sold?

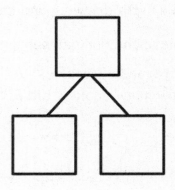

_____ shirts were sold.

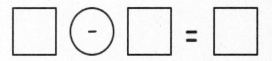

4. There were 9 children at the park. Some children went inside. Five children stayed.
   How many children went inside?

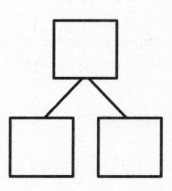

_____ children went inside.

Lesson 31:   Solve *take from with change unknown* math stories with drawings.

EUREKA
MATH

Name _____   Date _____

Make a math drawing, and circle the part you know. Cross out the unknown part.
Complete the number sentence and number bond.

Deb blows up 9 balloons. Some balloons popped. Three balloons are left.
How many balloons popped?

_____ balloons popped.

Read

There are 8 juice boxes in the cubbies. Some children drink their juice.
Now, there are only 5 juice boxes. How many juice boxes were taken from
the cubbies?

Make a number bond. Write a subtraction sentence and a statement to
match the story. Make a box around the solution in your number sentence.
Make a math drawing to show how you know.

Draw

Write

Lesson 32: Solve *put together/take apart with addend unknown* math stories.

Name _____ Date _____

Solve. Use simple math drawings to show how to solve with addition and subtraction.
Label the number bond.

1.

There are 5 apples.
Four are Sam's.
The rest are Jim's.
How many apples does Jim have?

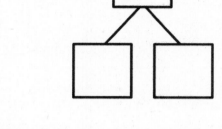

Jim has _____ apple.

$$\boxed{} + \boxed{} = \boxed{5}$$

$$\boxed{5} - \boxed{} = \boxed{}$$

2.

There are 8 mushrooms. Five are black. The rest are white.
How many mushrooms are white?

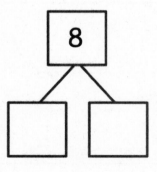

_____ mushrooms are white.

$$\boxed{} + \boxed{} = \boxed{8}$$

$$\boxed{8} - \boxed{} = \boxed{}$$

EUREKA MATH™

Lesson 32:   Solve *put together/take apart with addend unknown* math stories.

217

©2018 Great Minds®. eureka-math.org

Use the number bond to complete the number sentences. Use simple math drawings to tell math stories.

3.

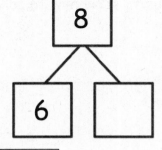

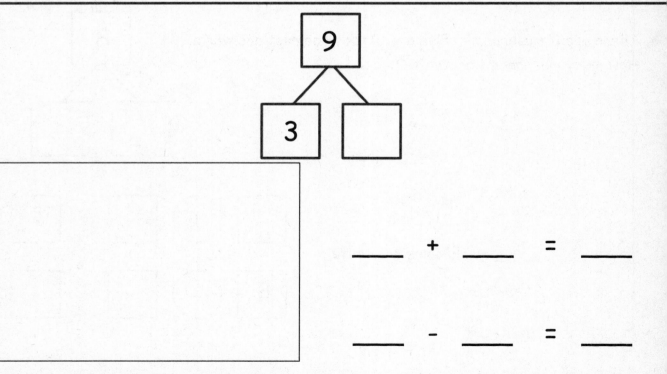

_____ + _____ = 8

8 - _____ = _____

4.

_____ + _____ = _____

_____ - _____ = _____

Lesson 32:   Solve *put together/take apart with addend unknown* math stories.

EUREKA MATH

Read

Nine children are playing outside. One child is on the swings and the rest are playing tag. How many children are playing tag?

Write a number bond and number sentence. Make a math drawing to show how you know.

Draw

Lesson 33:   Model 0 less and 1 less pictorially and as subtraction number
              sentences.

©2018 Great Minds®. eureka-math.org

221

Write

There are [ ] children playing tag.

Lesson 33:   Model 0 less and 1 less pictorially and as subtraction number sentences.

©2018 Great Minds®. eureka-math.org

EUREKA
MATH™

Name _____   Date _____

Cross off, when needed, to subtract.

1.      2.   

$6 - 1 =$ ___ $6 - 0 =$ ___

If you want, make a 5-group drawing for each problem like the ones above.
Show the subtraction.

3.   4.

$7 - 1 =$ ___ $7 - 0 =$ ___

5.   6.

$10 - 1 =$ ___ $10 - 0 =$ ___

7.   8.

$8 - 1 =$ ___ $8 - 0 =$ ___

9.   10.

$9 - 1 =$ ___ $9 - 0 =$ ___

Cross off, when needed, to subtract.

11.

   6 – 1 = ___

12.

   8 – 1 = ___

13.

   9 – 0 = ___

Subtract.

14.   7 – 1 = ___   15.   8 – 0 = ___   16.   9 – 1 = ___

17. Fill in the missing number.   Visualize your 5-groups to help you.

   a. 6 – 0 = ___   b. 6 – 1 = ___

   c. 7 – ___ = 7   d. 7 – 1 = ___

   e. 8 – 0 = ___   f. 8 – ___ = 7

   g. 9 – ___ = 9   h. 9 – 1 = ___

   i. 10 – ___ = 10   j. 10 – ___ = 9

Lesson 33:   Model 0 less and 1 less pictorially and as subtraction number
sentences.

©2018 Great Minds®. eureka-math.org

EUREKA MATH

Name _____ Date _____

Complete the number sentences. If you want, use 5-group drawings to show the subtraction.

1.

$9 - 1 =$ ____

2.

$8 =$ ____ $- 0$

3.

$8 =$ ____ $- 1$

4.

$10 = 10 -$ ____

EUREKA MATH™

Lesson 33:   Model 0 less and 1 less pictorially and as subtraction number sentences.

©2018 Great Minds®. eureka-math.org

225

Read

Eighty-three beads spill on the floor. A student picks up 1 bead. How many beads are still on the floor?

Write a number bond, number sentence, and a statement to share your solution.

Extension: If a second child picks up 10 more beads, how many beads will remain on the floor? Use number bonds to show how you know.

Draw

Write

_____

_____

_____

Lesson 34:   Model *n* − *n* and *n* − (*n* − 1) pictorially and as subtraction sentences.

EUREKA MATH

Name _____   Date _____

Cross off to subtract.

8-7 = _1_

1. ●●●●● ○

    6 – 6 = ___

2. ●●●●● ○

    6 – 5 = ___

Subtract. Make a math drawing, like those above, for each.

3.

    7 – 7 = ___

4.

    7 – 6 = ___

5.

    10 – 10 = ___

6.

    10 – 9 = ___

7.

    8 – 8 = ___

8.

    8 – 7 = ___

9.

    9 – 9 = ___

10.

    9 – 8 = ___

Cross off, when needed, to subtract.

11.

12.

13.

6 – 6 = ___

8 – 8 = ___

9 – 8 = ___

Subtract. Make a math drawing, like those above, for each.

14.

15.

16.

7 – 7 = ___

8 – 7 = ___

9 – 9 = ___

17. Fill in the missing number. Visualize your 5-groups to help you.

a. 6 – 6 = ___

b. 6 – 5 = ___

c. 7 – ___ = 0

d. 7 – 6 = ___

e. 8 – 8 = ___

f. 8 – ___ = 1

g. 9 – ___ = 0

h. 9 – 8 = ___

i. 10 – ___ = 10

j. 10 – ___ = 1

Lesson 34:   Model $n - n$ and $n - (n - 1)$ pictorially and as subtraction sentences.

EUREKA
MATH™

Name _____   Date _____

Make 5-group drawings to show the subtraction.

1.

$9 - \underline{} = 1$

2.

$0 = 10 - \underline{}$

3.

$1 = \underline{} - 7$

4.

$0 = \underline{} - 9$

Read

The teacher spilled 18 beads on the floor today. A student picked up 17 of the beads. How many beads are still left on the floor?

Write a number bond, number sentence, and a statement to share your solution.

Extension: If the 17 beads had been picked up by two students, how many beads might each student have picked up? Make a number bond to show your solution.

EUREKA MATH

Lesson 35:   Relate subtraction facts involving fives and doubles to corresponding decompositions.

233

©2018 Great Minds®. eureka-math.org

Draw

Write

_____

_____

_____

Lesson 35:   Relate subtraction facts involving fives and doubles to corresponding decompositions.

EUREKA MATH™

Name _____   Date _____

Solve the sets of number sentences. Look for easy groups to cross off.

1.

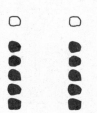

6 – 5 = ____

6 – 1 = ____

2.

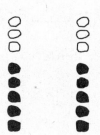

8 – 3 = ____

8 – 5 = ____

3.

9 – 4 = ____

9 – 5 = ____

Subtract. Make a math drawing for each problem like the ones above. Write a number bond.

4.

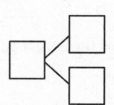

7 – 5 = ____

7 – 2 = ____

5.

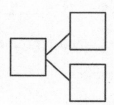

10 – 5 = ____

EUREKA MATH

Lesson 35:   Relate subtraction facts involving fives and doubles to corresponding decompositions.

©2018 Great Minds®. eureka-math.org

235

6. Solve. Visualize your 5-groups to help you.

a. 7 – 5 = ___

b. 7 – ___ = 5

c. 8 – 3 = ___

d. 9 – ___ = 4

e. 9 – ___ = 5

f. 8 – ___ = 3

Complete the number bond and number sentence for each problem.

7. 4 – 2 = ___

8. 6 – 3 = ___

9. 10 – 5 = ___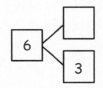

10. 8 – 4 = ___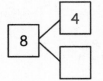

11. 8 – 4 = ___

12. 6 – 3 = ___

13. Complete the number sentences below. Circle the strategy that can help.

a. 7 – 5 = ___ 5-groups | doubles

b. 7 – 2 = ___ 5-groups | doubles

c. 8 – 4 = ___ 5-groups | doubles

d. 8 – 3 = ___ 5-groups | doubles

e. 8 – 5 = ___ 5-groups | doubles

f. 10 – 5 = ___ 5-groups | doubles

Lesson 35: Relate subtraction facts involving fives and doubles to corresponding decompositions.

©2018 Great Minds®. eureka-math.org

EUREKA MATH™

Name _____   Date _____

Solve the number sentences. Make a number bond.
Draw a picture or write a statement about the strategy that helped you.

Doubles helped me
solve!

$6 - 3 = 3$

1. _____ $- 5 = 5$ 2. $8 -$ _____ $= 4$ 3. $9 -$ ___ $= 4$

EUREKA MATH™   **Lesson 35:**   Relate subtraction facts involving fives and doubles to corresponding   **237**
decompositions.

©2018 Great Minds®. eureka-math.org

Read

There are 10 beads on the floor. There is the same number of red beads as white beads. A student picks up the white beads. How many beads are still on the floor?

Write a number bond, number sentence, and a statement to share your solution. Make a math drawing to show how you know.

Draw

Write

_____

_____

_____

Lesson 36:   Relate subtraction from 10 to corresponding decompositions.

Name _____   Date _____

Solve the sets. Cross off on the 5-groups.
Use the first number sentence to help you solve the next.

1.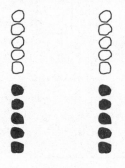

$$10 - 9 = \underline{\hspace{1cm}}$$

$$10 - 1 = \underline{\hspace{1cm}}$$

2.

$$10 - 6 = \underline{\hspace{1cm}}$$

$$10 - 4 = \underline{\hspace{1cm}}$$

3.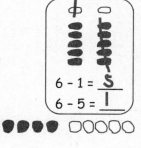

$$6 - 1 = \underline{5}$$
$$6 - 5 = \underline{1}$$

$$10 - 3 = \underline{\hspace{1cm}}$$

$$10 - 7 = \underline{\hspace{1cm}}$$

Make a math drawing and solve.

4.

$$10 - 4 = \underline{\hspace{1cm}}$$

$$10 - 6 = \underline{\hspace{1cm}}$$

5.

$$10 - 5 = \underline{\hspace{1cm}}$$

6.

$$10 - 8 = \underline{\hspace{1cm}}$$

$$10 - 2 = \underline{\hspace{1cm}}$$

Subtract. Then, write the related subtraction sentence.
Make a math drawing if needed, and complete a number bond for each.

7.

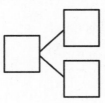

$$10 - 8 = \rule{1.5cm}{0.4pt}$$

\rule{10cm}{0.4pt}

8.

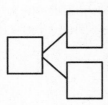

$$10 - 9 = \rule{1.5cm}{0.4pt}$$

\rule{10cm}{0.4pt}

9.

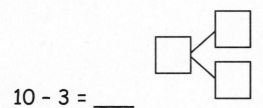

$$10 - 3 = \rule{1.5cm}{0.4pt}$$

\rule{10cm}{0.4pt}

10.

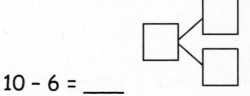

$$10 - 6 = \rule{1.5cm}{0.4pt}$$

\rule{10cm}{0.4pt}

11. Fill in the missing part. Write the 2 matching subtraction sentences.

a.

b.

c.
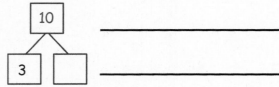

d.
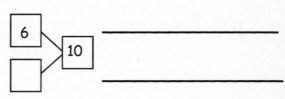

e.
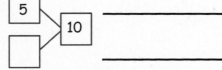

Lesson 36:   Relate subtraction from 10 to corresponding decompositions.   EUREKA MATH™

Name _____   Date _____

Fill in the missing part. Draw a math picture if needed. Write the 2 matching subtraction sentences.

1.

_____

_____

2.

_____

_____

3.

10

4

_____

_____

Read

There are 10 beads on the floor. A student picked up some of the beads but left some on the floor. Write a number bond and a number sentence that would match this story.

Extension: What other number bonds and number sentences could match this story? Try to list all of the possibilities.

Draw

Write

_____

_____

_____

Lesson 37: Relate subtraction from 9 to corresponding decompositions.

EUREKA MATH

Name _____   Date _____

Solve the sets. Cross off on the 5-groups. Write the related subtraction sentence that would have the same number bond.

1.

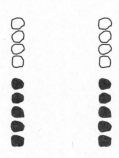

$9 - 8 =$ ___

$9 - 1 =$ ___

2.

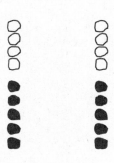

$9 - 7 =$ ___

_____

3.

$9 - 9 =$ ___

_____

Make a 5-group drawing. Solve, and write a related subtraction sentence that would have the same number bond. Cross off to show.

4.

$9 - 6 =$ ___

_____

5.

$9 - 4 =$ ___

_____

6.

$9 - 3 =$ ___

_____

Subtract. Then, write the related subtraction sentence.
Make a math drawing if needed, and complete a number bond.

7.

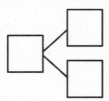

  9 – 5 = ___

  _____

8.

  9 – 8 = ___

  _____

9.

  9 – 7 = ___

  _____

10.

  9 – 3 = ___

  _____

11. Fill in the missing part. Write the 2 matching subtraction sentences.

a. 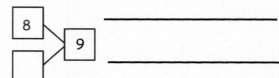   _____
                          _____

b. _____
    _____

c. 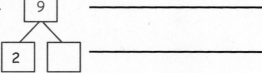   _____
                          _____

d. 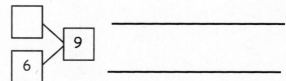   _____
                          _____

e. 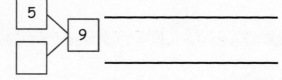   _____
                          _____

EUREKA
MATH™

Name _____   Date _____

Fill in the missing part. Draw a math picture if needed. Write the 2 matching subtraction sentences.

1.

_____

_____

2.

_____

_____

3.

_____

_____

Read

Jessie and Carl were comparing the beads they picked up. Jessie picked up 9 beads. 5 of them were red, and the rest were white. Carl picked up

5 red beads and 4 white beads. Carl said they had the same number of white beads. Is Carl correct?

Draw and label your work to show your thinking.

Draw

Lesson 38:   Look for and make use of repeated reasoning and structure using the
addition chart to solve subtraction problems.

251

©2018 Great Minds®. eureka-math.org

Write

_____

_____

_____

Lesson 38: Look for and make use of repeated reasoning and structure using the addition chart to solve subtraction problems.

©2018 Great Minds®. eureka-math.org

EUREKA
MATH™

Name _____ Date _____

| 6 – 4 |

Pick a subtraction card.

Find the related addition fact on the chart and shade it in.

Write the subtraction sentence and a number bond to match.

Continue for at least 6 turns.

1+9									
1+8	2+8								
1+7	2+7	3+7							
1+6	2+6	3+6	4+6						
1+5	2+5	3+5	4+5	5+5					
1+4	2+4	3+4	4+4	5+4	6+4				
1+3	2+3	3+3	4+3	5+3	6+3	7+3			
1+2	2+2	3+2	4+2	5+2	6+2	7+2	8+2		
1+1	2+1	3+1	4+1	5+1	6+1	7+1	8+1	9+1	
1+0	2+0	3+0	4+0	5+0	6+0	7+0	8+0	9+0	10+0

Lesson 38: Look for and make use of repeated reasoning and structure using the addition chart to solve subtraction problems.

253

EUREKA MATH™

©2018 Great Minds®. eureka-math.org

On your addition chart, shade a square orange. Write the related subtraction fact in a space below with its number bond. Color all the totals orange.

1. _____ - _____ = _____

2. _____ - _____ = _____

3. _____ - _____ = _____

4. _____ = _____ - _____

5. _____ = _____ - _____

Lesson 38: Look for and make use of repeated reasoning and structure using the addition chart to solve subtraction problems.

255

©2018 Great Minds®. eureka-math.org

EUREKA MATH

Name _____   Date _____

Write the related number sentences for the number bonds.

1.

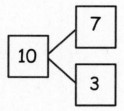

2.

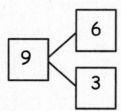

_____ - _____ = _____   _____ - _____ = _____

_____ + _____ = _____   _____ + _____ = _____

_____ ◯ _____ = _____   _____ ◯ _____ = _____

_____ ◯ _____ = _____   _____ ◯ _____ = _____

EUREKA MATH   **Lesson 38:**   Look for and make use of repeated reasoning and structure using the addition chart to solve subtraction problems.   **257**

©2018 Great Minds®. eureka-math.org

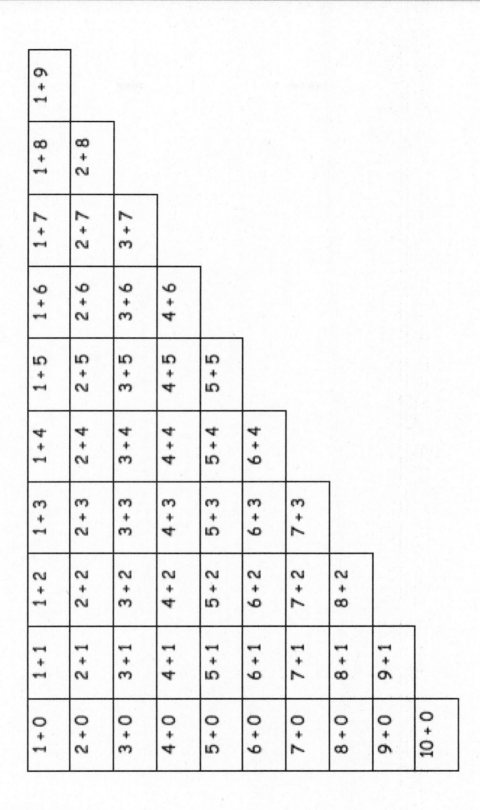

addition chart; from Lesson 21

Lesson 38: Look for and make use of repeated reasoning and structure using the addition chart to solve subtraction problems.

Read

John has 10 pencils. Mark has 9 pencils. Anna has 8 pencils. They each lost two of their pencils. How many do they each have now? Write a number bond and number sentence for each student.

Draw

Lesson 39:   Analyze the addition chart to create sets of related addition and subtraction facts.

©2018 Great Minds®. eureka-math.org

261

Write

_____

_____

_____

Lesson 39:   Analyze the addition chart to create sets of related addition and subtraction facts.

©2018 Great Minds®. eureka-math.org

EUREKA
MATH™

Name _____   Date _____

Study the addition chart to solve and write related problems.

1 + 9								
1 + 8	2 + 8							
1 + 7	2 + 7	3 + 7						
1 + 6	2 + 6	3 + 6	4 + 6					
1 + 5	2 + 5	3 + 5	4 + 5	5 + 5				
1 + 4	2 + 4	3 + 4	4 + 4	5 + 4	6 + 4			
1 + 3	2 + 3	3 + 3	4 + 3	5 + 3	6 + 3	7 + 3		
1 + 2	2 + 2	3 + 2	4 + 2	5 + 2	6 + 2	7 + 2	8 + 2	
1 + 1	2 + 1	3 + 1	4 + 1	5 + 1	6 + 1	7 + 1	8 + 1	9 + 1
1 + 0	2 + 0	3 + 0	4 + 0	5 + 0	6 + 0	7 + 0	8 + 0	9 + 0	10 + 0

Pick a subtraction card.

Find the related addition fact on the chart and shade it in.

Write the subtraction sentence and the shaded addition sentence.

Write the other two related facts.

Continue for at least 4 turns.

Lesson 39:   Analyze the addition chart to create sets of related addition and subtraction facts.

©2018 Great Minds®. eureka-math.org

263

Choose an expression card, and write 4 problems that use the same parts and totals. Shade the totals orange.

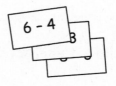

6 . 4 . 2
4 . 2 . 6
2 ⊕ 4 . 6
6 ⊖ 2 . 4

1. ____ - ____ = ____

    ____ + ____ = ____

    ____ 〇 ____ = ____

    ____ 〇 ____ = ____

2. ____ - ____ = ____

    ____ + ____ = ____

    ____ 〇 ____ = ____

    ____ 〇 ____ = ____

3. ____ - ____ = ____

    ____ + ____ = ____

    ____ 〇 ____ = ____

    ____ 〇 ____ = ____

4. ____ - ____ = ____

    ____ + ____ = ____

    ____ 〇 ____ = ____

    ____ 〇 ____ = ____

Lesson 39:   Analyze the addition chart to create sets of related addition and subtraction facts.

EUREKA MATH™

Name _____   Date _____

Write the related number sentences for the number bonds.

1.

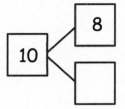

_____ - _____ = _____

_____ + _____ = _____

_____ ◯ _____ = _____

_____ ◯ _____ = _____

2.

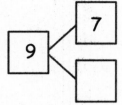

_____ - _____ = _____

_____ + _____ = _____

_____ ◯ _____ = _____

_____ ◯ _____ = _____

EUREKA MATH™

Lesson 39:   Analyze the addition chart to create sets of related addition and subtraction facts.

265

1+9									
1+8	2+8								
1+7	2+7	3+7							
1+6	2+6	3+6	4+6						
1+5	2+5	3+5	4+5	5+5					
1+4	2+4	3+4	4+4	5+4	6+4				
1+3	2+3	3+3	4+3	5+3	6+3	7+3			
1+2	2+2	3+2	4+2	5+2	6+2	7+2	8+2		
1+1	2+1	3+1	4+1	5+1	6+1	7+1	8+1	9+1	
1+0	2+0	3+0	4+0	5+0	6+0	7+0	8+0	9+0	10+0

addition chart; from Lesson 21

EUREKA MATH™   Lesson 39:   Analyze the addition chart to create sets of related addition and   267
                                                       subtraction facts.

©2018 Great Minds®. eureka-math.org

Credits

Great Minds® has made every effort to obtain permission for the reprinting of all copyrighted material. If any owner of copyrighted material is not acknowledged herein, please contact Great Minds for proper acknowledgment in all future editions and reprints of this module.